AF764959

Laser Scanning, Theory and Applications

Laser Scanning, Theory and Applications

Editor

Yijun Kwan

scitus academics

Laser Scanning, Theory and Applications
Edited by **Yijun Kwan**

ISBN: 978-1-68117-499-0
Library of Congress Control Number: 2016942841

© 2017 by
SCITUS Academics LLC,
www.scitusacademics.com
Box No. 4766, 616 Corporate Way,
Suite 2, Valley Cottage,
NY 10989

This book contains information obtained from highly regarded resources. Copyright for individual articles remains with the authors as indicated. All chapters are distributed under the terms of the Creative Commons Attribution License, which permits unrestricted use, distribution, and reproduction in any medium, provided the original author and source are credited.

Notice

Reasonable efforts have been made to publish reliable data and views articulated in the chapters are those of the individual contributors, and not necessarily those of the editors or publishers. Editors or publishers are not responsible for the accuracy of the information in the published chapters or consequences of their use. The publisher believes no responsibility for any damage or grievance to the persons or property arising out of the use of any materials, instructions, methods or thoughts in the book. The editors and the publisher have attempted to trace the copyright holders of all material reproduced in this publication and apologize to copyright holders if permission has not been obtained. If any copyright holder has not been acknowledged, please write to us so we may rectify.

Printed in United States of America on Acid Free Paper

Preface

The phrase laser scanning is used in contemporary engineering with two related but separate meanings. The first, much more basic meaning is the controlled, invisible or visible deflection of the laser beams. Scanned laser beams are used in certain 3D printers, in rapid prototyping, in material processing devices, in laser beam gravure machines, in ophthalmological laser devices due to presbyopia healing, in confocal microscopy, in laser printers, in laser shows, in ceramic laser treatments and also in barcode scanners. 3D Laser scanning is a non-contact, non-destructive technological innovation that captures the shape of physical objects digitally by using a laser light type. 3D laser beam scanners lead to point information clouds from an object's surface area. Simply put, 3D laser scanning is a way of capturing the actual shape and size of an object in the pc world as a 3-dimensional digital representation. 3D laser beam scanners evaluate fine details and obtain free shapes to create extremely precise point clouds easily. 3D item scanning enables the design process to be improved, speeds up and reduces errors in the compilation of information, saves money and time and thus makes it a stylish substitute for traditional information compilation techniques. 3D scanning is also used for mobile mapping, surveying, structural inspection and creation of interiors, as well as in archaeology.

Highlights the current developments in laser scanning, a highly powerful technique used in detection and measurement of features; the book explores fundamental work in theory of laser scanning or applies the principles of laser scanning to tackle medical, geodesic, biological and archaeological problems.

Table of Contents

CHAPTER 1

An Improved Method for Power-Line Reconstruction from Point Cloud Data

Bo Guo [1,2], Qingquan Li [1,*], Xianfeng Huang [3,4] and Chisheng Wang [1,2]

[1] Key Laboratory for Geo-Environment Monitoring of Coastal Zone of the National Administration of Surveying, Mapping and GeoInformation & Shenzhen Key Laboratory of Spatial Smart Sensing and Services, Shenzhen University, Nanhai Road 3688, Shenzhen 518060, China

[2] College of Civil Engineering, Shenzhen University, Nanhai Road 3688, Shenzhen 518060, China

[3] The State Key Laboratory of Information Engineering in Surveying, Mapping and Remote Sensing, Wuhan University, Wuhan 430072, China

[4] Collaborative Innovation Center of Geospatial Technology, Wuhan University, Wuhan 430072, China

ABSTRACT

This paper presents a robust algorithm to reconstruct power-lines using ALS technology. Point cloud data are automatically classified into five target classes before reconstruction. In order to improve upon the defaults of only using the local shape properties of a single power-line span in traditional methods, the distribution properties of power-line group between two neighbor pylons and contextual information of related pylon objects are used to improve the reconstruction results. First, the distribution properties of power-line sets are detected using a similarity detection method. Based on the probability of neighbor points belonging to the same span, a RANSAC rule based algorithm is then introduced to reconstruct power-lines through two important advancements: reliable initial parameters fitting and efficient candidate sample detection. Our experiments indicate that the proposed method is effective for reconstruction of power-lines from complex scenarios.

KEYWORDS

airborne laser scanning; power-line span; pylon; reconstruction

1. INTRODUCTION

The safety of power-line infrastructure significantly affects our everyday life and industrial activities. Automating the monitoring of high-voltage transmission lines is of importance to power utility companies [1]. There are two components of the automated monitoring methodology: detection of potential hazards, such as encroaching vegetation [2,3], and the analysis of power-line structural stability. Therefore, an important consideration when reconstructing 3D power-lines is to assert that the physical parameters of the cable are still within a safe margin while quickly detecting anomalies or defects [4].

Due to its ability to provide dense, fast and accurate measurements, Airborne Laser Scanning (ALS) is becoming increasingly popular for extensive, large area surveying tasks [5,6]. Used in power-line settings, ALS is able to extract information about objects with respect to power-line network system [7,8]. Therefore, an algorithm to reconstruct individual spans using ALS is introduced in this paper to model the sag of individual spans, or to compute its parameters.

Before reconstruction, the point cloud data have to be classified and extract the power-line points. Axelsson [9] introduced a method of three steps to extracting power-line points: (1) separation of ground surface; (2) classification of objects in power-lines and vegetation; and (3) refining classification of power-lines. Clode and Rottensteiner [10] detected trees and power lines from less than one point per m^2 point cloud in Sydney. First and last pulse return differences were applied. Zhu and Hyyppä proposed an automated power line detection method for forest environments [11].

There have been many methods for power-line span reconstruction, which mainly focus on constructing a singular line span. McLaughlin defined the parametric model of a power-line span, and extracted each individual span from classified power-line point cloud data [12]. Firstly, the model is fitted locally to a small seed region, which is then grown by adding neighbor points consistent with the trend of the model. After a region has been grown, the parameters of its corresponding model are re-estimated. This process is iterated until a given termination criteria is met. Jwa et al., introduced a method for detecting and modeling power-line spans based on perceptual grouping framework which groups similar features at different levels of information using a Voxel-based Piece-wise Line Detector (VPLD) method for power-line modeling [13]. Melzer and Briese firstly use a 2D Hough transformation to locate the groups of parallel power-lines in the projection points onto the XY plane. A 3D fit for each power-line span is then computed locally within its corresponding corridor [14]. Liang et al. considered that points in the same power-line would be close together whereas the points in different power-lines would be far apart [15]. Therefore, a power-line points clustering method based on KD trees was adopted. This method only considered the point distance and it ignored breakage points, which may have detrimental effects. Jwa and Sohn proposed a piecewise catenary curve model-growing algorithm to identify the points, which achieved clustering by iterative catenary curve fitting and cubic growing [16]. To a certain degree, this method is limited with the fitting curve and the selection of the initial growing cube. Based on the piecewise model growing [16], Sohn et al., precisely modeled power-line spans with catenary curve models in 3D, once the pylon localization is accomplished [17]. Cheng et al. introduced a method based on vehicle-borne LiDAR to cluster of power-line points, in which the initial clustering and clustering recovery procedures were conducted iteratively to identify each power line [18]. Also the sag reconstruction results are mainly based on robustness of the methods, it is also influenced to ambient conditions such as the temperature and the ageing of spans [19]. Lu and Kieloch [20] surveyed the modeling accuracy of power lines based on conductor temperature. They found that wind, electrical load and solar radiation are the three major factors that may affect the modeling accuracy. Because the data used in this paper were mainly collected in one time, we mainly focused on the reconstruction on certain ambient conditions when the point cloud data were obtained.

The methods mentioned above provide good results for power-lines reconstruction. However, if data omission occurs due to variations of laser position, occlusions, and insufficient data resolution, there are many failures such as undetected and splitting of transmission lines [12]. Therefore, more robust and integrated power-line modeling is important for monitoring of power safety.

Power-lines are distributed regularly between two neighbor pylons and are nearly parallel, with the same direction, sag and other physical properties. Using all of this information can improve existing power-line modeling results which only consider local information of power-line spans.

1) Power-lines and pylons are two types of related objects. In the power-line reconstruction process, we can use the distribution of neighbor pylons to improve the reconstruction results of power-lines.

2) The group set of power-lines spanning two pylons have same properties. Thus, the parameters of each line span are similar. Therefore, reconstructing a group set of power-line spans between two neighbor pylons together can avoid the fitting failures of initial parameters by reconstructing power-line span one by one.

In this paper, an improved and robust algorithm is proposed to reconstruct transmission line spans from ALS data based on the two properties mentioned above, in order to overcome the reconstruction failures caused by data sparseness and occlusion. Before reconstruction, data are firstly classified into five categories: power-line, pylon, vegetation, building and ground. Then, an algorithm based on RANdom SAmple Consensus (RANSAC) rule [21] is introduced to reconstruct each individual power-line span.

2. POINT CLOUD DATA CLASSIFICATION

In this section, an algorithm is proposed to classify data into five categories: power-line, vegetation, building, ground and pylon. This classification process is mainly based on our previous work [22], which is a machine learning [23,24] algorithm. Firstly, 26 features based on the geometry and echo information of the point cloud data are calculated. A JointBoost classifier [25] is then used to classify the point cloud data into different target classes. The objects in five categories are classified simultaneously rather than by directly extracing power-line points. Although direct power-line extraction methods using discriminant rules [14] are effective, these methods still have inherent limitations when dealing with complex landscapes, and they lack the contextual information [26,27] provided by other kinds of objects. The simultaneous classification can be used to overcome these defaults, and as we mention before, the detection of pylons together is helpful for power-lines reconstruction. As for power-line point detection, the features, which can effectively distinguish line objects, such as line echo, linearity and Hough transformation are mainly used [22].

In the JointBoost classifier training step, balanced object-based sampling is important [28]. In our study, the training step was performed with samples (points with true labels) of equal number per class. Another part of the training set was randomly chosen as a validation set. The validation set was used to measure the effectiveness of the classifier parameters. The trained classifier was then used to classify the test data. A well-trained JointBoost classifier with properly selected features can generate good classification results [22]. The precision and recall of the classification are important considerations for power-line reconstruction. It is found that both the precision and recall for the pylon and power-lines are high (Section 4.1).

3. 3D RECONSTRUCTION OF POWER-LINE SPANS

A power-line span can be reconstructed using a certain pattern which can be represented by a catenary curve [12]. The traditional power-line reconstruction are mainly region grouping based [12,13]. Using these methods to estimate the parameters of a line span, the algorithms usually begin with an arbitrarily chosen local section, and then iteratively identifying adjacent samples on the same span, progressively redefining the catenary parameters. The extraction and reconstruction of some singular span could fail in some situations: splitting, undetected, etc. This is primarily due to sparseness of the data, resulting in a failure to first compute a reliable local model or being unable to find the connected samples.

In order to overcome these faults, we introduce an improved method based on RANSAC rule, which is an iterative method to estimate the parameters of a mathematical model from a set of observed data that contains a number of outliers, resulting in robustly reconstructed power-line spans. During the iterations, general RANSAC generates a large number of hypotheses of the mathematical model parameters by randomly selecting a minimum subset of samples that each uniquely determines a set of parameters of the model. After determining an initial hypothesis of the mathematical model, the other samples are gradually added into the model to determine whether they are inliers or outliers. At last, the final results are determined if their number of inliers selected by RANSAC is bigger than a voting threshold [29,30]. Using the RANSAC rule in power-line reconstruction can avoid failures of initial parameters fitting which caused by the sparseness of data.

There are two key concepts in RANSAC: (1) setting the initial parameters and (2) searching for candidate samples. The inherent information of power-line distribution properties can be used to modify these two concepts in the traditional RANSAC rule in order to improve the efficiency and precision of reconstruction.

Based on the parallel distribution of power-line spans, we first use a segmentation method (Section 3.1) to detect the similar distribution of a group of power-line spans between two neighbor pylons and further to roughly detect points that belong to a power-line span with high probability. We then improve upon the two key concepts in RANSAC rule by using this rough detection (Section 3.2).

3.1. Similarity Detection

The distribution of power-line group set between two neighbor pylons are similar: roughly parallel, with the same sag and direction. A profile segmentation based method is used to detect these similar properties. Figure 1 outlines the method of segmentation. Firstly, the power-line point data between two neighbor pylons are segmented into profiles (indicated with transparent cuboids) equally distant along the main direction of the power-lines. Then, the points within a profile are projected to its main plane (indicated with a yellow plane). A region growing algorithm is used to connect the comparable projecting points and create groups (indicated in Figure 1, a number represents a center point of group). This segmentation method are also used for constructing neighborhood space of points. There are three levels for neighborhood space construction: point-based level [31], voxel-based level [32,33] and object-level [34]. The points within a group of a profile are considered to be within a neighborhood which containing local structure of a singular power-line. In order to maintain consistency of the coordinates among the different profiles, the coordinates for a projection plane are constructed by the following rules: (1) the X and Y axes are orthogonal to the main direction of the power-lines; and (2) the point of origin is a certain point projecting to this projection plane. As shown in Figure 1, projected center points on neighboring profiles have a similar distribution.

The similar properties are used to roughly detect the points belonging to the same power-line among different profiles. We first explain this method between two profiles (Section 3.1.1), and then expand it to many profiles (Section 3.1.2).

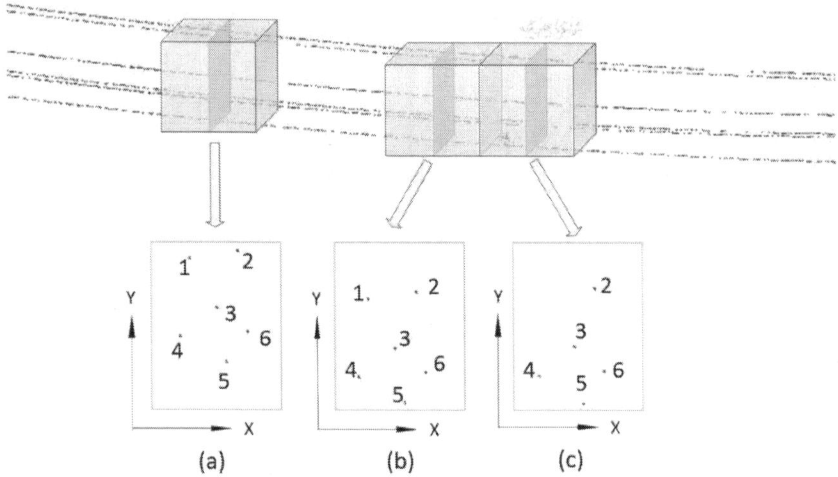

Figure 1. Similarity detection by segmentation. The corresponding center points among different profiles (represented by (**a**–**c**)) for the same power-line span are given the same number. In instances of missing data, the points do not correspond.

3.1.1. Similarity Detection between Two Neighbor Profiles

Given two corresponding points p i and p j on two neighbor profiles P i and P j , respectively, we use two principles, neighboring and geometry matching, to detect how probable these two points are belonging to the same power-line span.

The Neighboring Ratio

Due to the projection way, the locations of center points of each power-line on adjacent profiles are similar (see Figure 1b,c). Thus, we first use distance to test for points on a specific power-line span between two neighbor projecting profiles. The distance D is defined as $D = \sqrt{(x_i - x_j)^2 + (y_i - y_j)^2}$, where (x i , x j) and (y i , y j) are the coordinates of points p i and p j on profiles P i and P j , respectively.

The smaller this distance, the more probable that the corresponding points belong to a singular power-line span. The neighboring ratio is defined as:

$$R_n = 1 - \frac{D}{D_{\max}} \quad (1)$$

where, D max is the largest value of D .

The Geometry Matching Ratio

As mentioned previously, because of omitting or outlying power-line points, there may be failures to connect points only using the neighboring ratio. For this reason, we additionally introduce a geometry matching based method to detect the geometric similarity of the layout of points on adjacent profiles.

Due to the parallel property of power-line group, there must be some structural similarities between the projecting points of two profiles. Therefore, using geometry matching methods [35] is helpful for identifying the points belonging to a power-line span between different profiles. In order to overcome the problems of noise and missing points, we use similar triangles based point set matching method [36,37,38]. This method to detect similarity is independent of the coordinates and transformation (Figure 2).

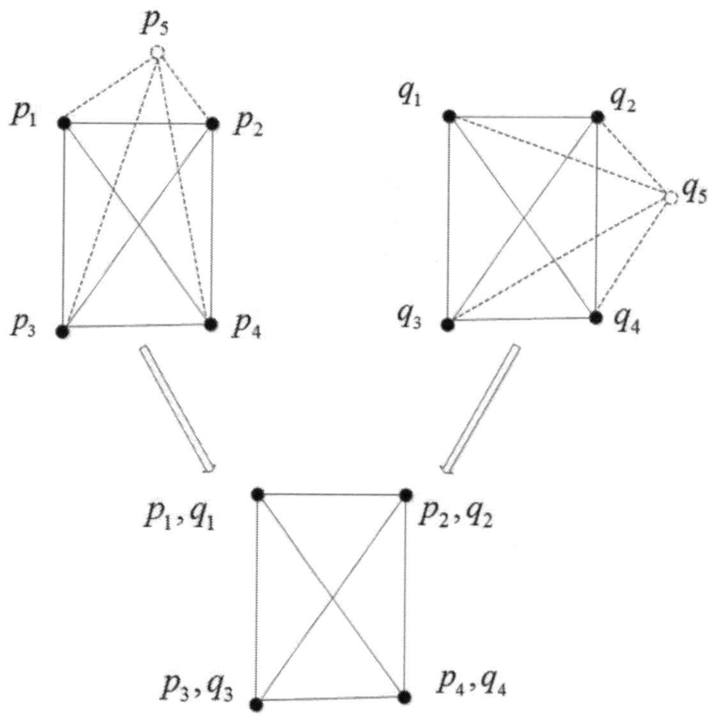

Figure 2. Geometry matching based on similar triangles. Based on the frequency that triangles match, we can easily get the corresponding points and conclude that p 5 and q 5 are noise or undetected points.

Two triangle sets are created from the triplets of points in each profile. In geometry, similar triangles have the following properties: the corresponding interior angles are equal and corresponding edges are proportional. The triangles from the two triangle sets are searched to find all of the possible similar pairs and the most frequently matched vertices are treated as corresponding points. In Figure 2, if p i matches q j most frequently, it can be concluded that these are corresponding points and each belong to a power-line span with high probability.

Let us assume p i matches q j with N number, the geometry matching ratio is defined as:

$$R_g = \frac{N}{N_{max}} \tag{2}$$

where N max is the maximum number of N .

3.1.2. Similarity Detection among Many Profiles

Rather than only using two neighbor profiles, we expand the similarity detection method by comparing one profile with many other profiles within a range (Figure 3). The range is defined by the biggest gap between successive points on a line and the vertical distance of power-line spans.

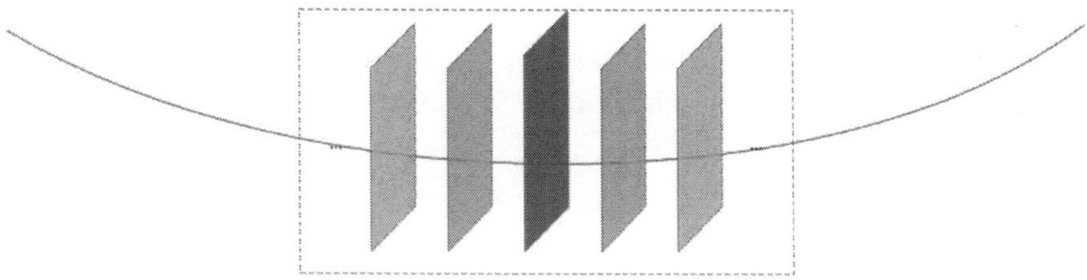

Figure 3. A main profile matches with many others within a range (indicated using the rectangle of dotted lines). The main profile is displayed with red plane, other matched profiles are displayed with yellow planes.

3.1.3. Seed Section Detection and Candidate Samples Searching

The similarity can be used to roughly detect points belong to the same spans, and therefore is mainly used to detect seed sections of power-line spans, and search candidate samples.

Seed Section Detection

Before the reconstruction of a power-line span, we have to search a seed section on which all the points must belong to this span. Given a section, we suppose its center points p_1, p_2, $\cdots$ p_n exist, respectively, on profiles P_I, P_{II}, $\cdots$ P_N (Figure 4).

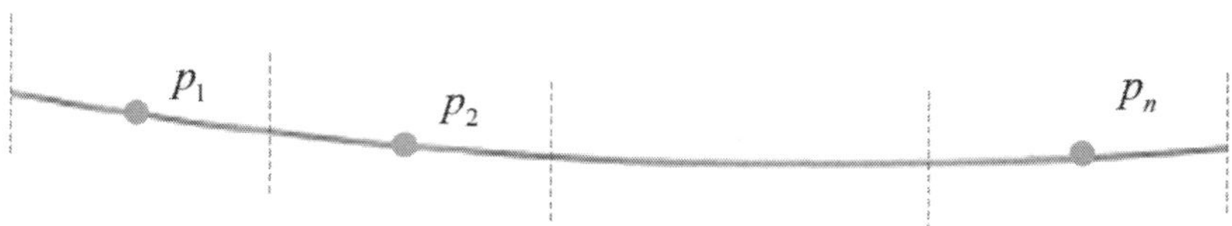

Figure 4. p_1, p_2, $\cdots$ p_n exist, respectively, on profiles P_I, P_{II}, $\cdots$ P_N on a seed section.

We use a chain rule to detect all points on this seed section. Arbitrarily chose a points p_i on profile P_I, there must be always at least one point p_j on other profiles P_J on this seed section meet the strict conditions: the neighboring and geometry matching ratio are both bigger than its defined threshold.

$$\left.\begin{array}{l} R_n > \delta_n \\ R_g > \delta_g \end{array}\right\} \tag{3}$$

where R_n is the neighboring ratio, R_g is the geometry matching ratio, δ_n and δ_g are neighboring and geometry matching threshold, respectively.

It is ensured that all points on this section are tightly connected by these strict conditions.

Candidate Sample Detection

During the reconstruction process, given a point p i on profile P I on a reconstructed section, we chose a corresponding point p j on neighbor profile P J to measure the probability that the two points belong to a same power-line span. The probability value is defined as:

$$R = R_n \times R_g \tag{4}$$

where, R n is the neighboring ratio, and R g is the geometry matching ratio.

3.2. RANSAC Based Reconstruction

After initially detecting points on a power-line span using similarity detection (Section 3.1), a RANSAC rule is designed for precisely reconstructing power-line spans between two neighbor pylons. As mentioned previously, the RANSAC rule has two key parts: setting the initial parameters and searching for candidate samples.

3.2.1. Setting the Initial Parameters

Given a seed section, we have to set its initial parameters. Due to the parallel nature of power-lines between two neighbor pylons, the parameters of some power-line spans in a group may be same. Therefore, using the parameters of properly fitted power-line spans as initial parameters to others for reconstructing is helpful for avoiding the failure of initial parameters fitting.

If there is no properly initial parameters searched, we have to select a seed section and determine its initial parameters according to [12]. The similarity detection is used to identify suitable seed sections (Section 3.1). It has been found that as the length of the seed section increases, the rate of successful reconstruction will be higher. Therefore, the sections with bigger length are first tested and then those with reducing length are sequentially selected as seeds.

3.2.2. Candidate Sample Detection

After seed section selecting and initial parameters fitting, in the RANSAC reconstruction process, candidate samples are added by selecting neighbor profile center points with high corresponding probability with the reconstructed section. The increasing power-line span is fitted using numerical methods [39] to check whether the added samples could meet a convergence rule: whether the residual of fitted function to data is smaller than a threshold. If not, the candidate points with decreasing probability are sequentially added to be checked until converging points are found, or there are no points remaining.

Due to these two favorable properties, the RANSAC method can improve reconstruction results. Figure 5 describes the flow chart of RANSAC for power-line span reconstruction between two connected neighboring pylons. At the beginning, we select a seed section with similarity detection. The other candidate samples are gradually added into the model to determine whether they are inliers or outliers. After incorporating all of the inlier points, the parameters of the power-line span are recalculated. These steps run iteratively until the parameters converge. The power-line span model is then chosen as the final result if their length is bigger than a defined threshold. After all spans between two connected neighboring pylons are iteratively reconstructed using the rules described above, the final output is all well-reconstructed spans between these two neighbor pylons.

4. EXPERIMENTAL ANALYSIS AND RESULTS

The results were obtained using a dataset that has been acquired for power-line management in Hubei, China. The data were collected using the Riegl Q560. Due to the topography, the flying altitude varied significantly and was approximately 200–600 m above the ground. The scan angle was $\pm$ 22.5 ° . It was swept across an average swath of width 160–500 m. Based on the changing of flying height and speed, the point density variation was large. The gap between successive points was 0.2–1.2 m. Due to data omission and misclassification, there were also many large gaps up to 10 m.

Approximately 20 km of data were obtained. The dataset was partitioned into two subsets for testing and training. The testing dataset comprise around 412 power-line spans. Testing data were manually labeled as classes of ground truth. Additionally, each individual power-line was identified for analyzing the reconstruction. In order to be clearly demonstrated, we mainly show typical reconstruction results in this section.

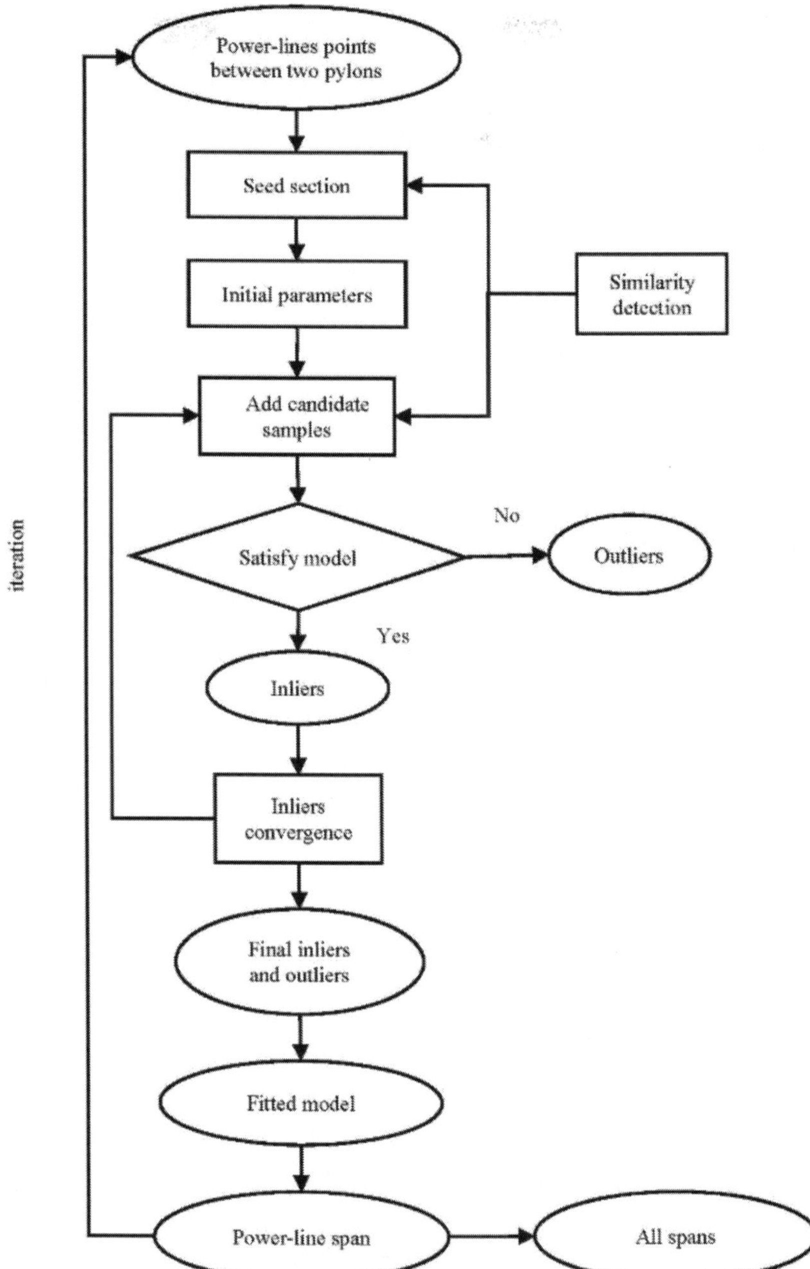

Figure 5. Progress of RANSAC rule for power-line spans reconstruction.

We first classified the data, and then extracted the power-line and pylon points. The classification accuracy is important for objects reconstruction. Therefore we have to analyze the precision and recall of classification and eliminate errors. After classification, the power-lines were reconstructed using an improved RANSAC method. In order to analyze the robustness of our methods, the reconstruction results obtained using different methods were compared.

4.1. Initial Classification

In classification, an algorithm is proposed to classify data into the categories: power-lines, vegetation, ground, pylon, and building. The training step is performed with 4000 samples (points with true labels) per class except pylons and power-lines, randomly selected from the training dataset. The training samples for power-lines and pylons are twice the size as each of the other classes in order to improve the extraction accuracy of these two kinds of objects. Typical results are shown in Figure 6.

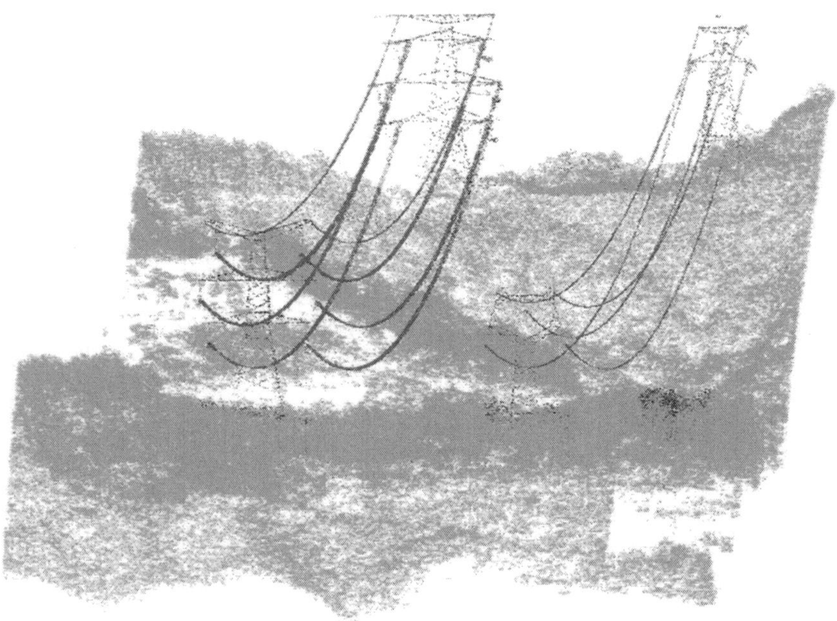

Figure 6. Classification results. For simplification, the ground and building points are not shown in this figure. (color legend: trees, green; power-lines, violet; and pylons, blue).

The classes of major points are correctly identified using the classification algorithm [21]. However, there are many misclassifications. Some points representing power-lines adjacent to the region of pylons are misclassified as vegetation or pylon. Some points representing vegetation are misclassified as power-lines because of the sparse distribution. Misclassified points are also detected in pylons where they are labeled as power-lines. The main reason is that the structure of a pylon is comprised of multiple metallic triangles for stabilization, and the sides of the triangles share the same linear structure characteristics as power-lines. There are also many false vegetation points that belong to pylons. This is because the densities of pylon points vary in different regions, especially the intersecting regions between power-lines and pylons and regions near the ground.

The precision and recall of power-line and pylon are shown in Table 1.The situation of some vegetation being misclassified as pylons is harmful. In order to precisely use the information of pylons' distribution, we have to eliminate the false positive. Between two neighbor pylons, a Hough-Transformation (HT) is used to detect whether they are connected by power-lines [14]. Detected ALS power-line points are firstly projected onto a XY plane and apply the HT to measure a maximum voting score accumulated in HT space as indicating a degree of the presence of linear feature. If the maximum voting score is corresponding to the angle of connecting line by these two neighbor pylons, these pylons are true positive. Otherwise, either of these pylon may be false positive.

Due to the effectiveness of classification process, although there are many errors, the majority of points are correctly classified.

The situation of some vegetation being misclassified as pylons is harmful. In order to precisely use the information of pylons' distribution, we have to eliminate the false positive. Between two neighbor pylons, a Hough-Transformation (HT) is used to detect whether they are connected by power-lines [14]. Detected ALS power-line points are firstly projected onto a XY plane and apply the HT to measure a maximum voting score accumulated in HT space as indicating a degree of the presence of linear feature. If the maximum voting score is corresponding to the angle of connecting line by these two neighbor pylons, these pylons are true positive. Otherwise, either of these pylon may be false positive.

4.2. Power-Lines Reconstruction

In this section, the robust method is proposed to extract and estimate the parameters of each individual span of power-line. We first use the similarity detection rule to initially detect the points belong to each same power-line span section. Based on the seed section detecting and candidate samples searching by the similarity, the RANSAC rule is used to robustly reconstruct the power-line spans. When the data quality is high, the reconstruction is precise and robust (Figure 7).

Table 1. Classification accuracy.

Overall Accuracy: 94.9%					
–	Ground	Vegetation	Building	Power-Line	Pylon
Precision	0.96	0.92	0.91	0.89	0.93
Recall	0.98	0.88	0.89	0.86	0.78

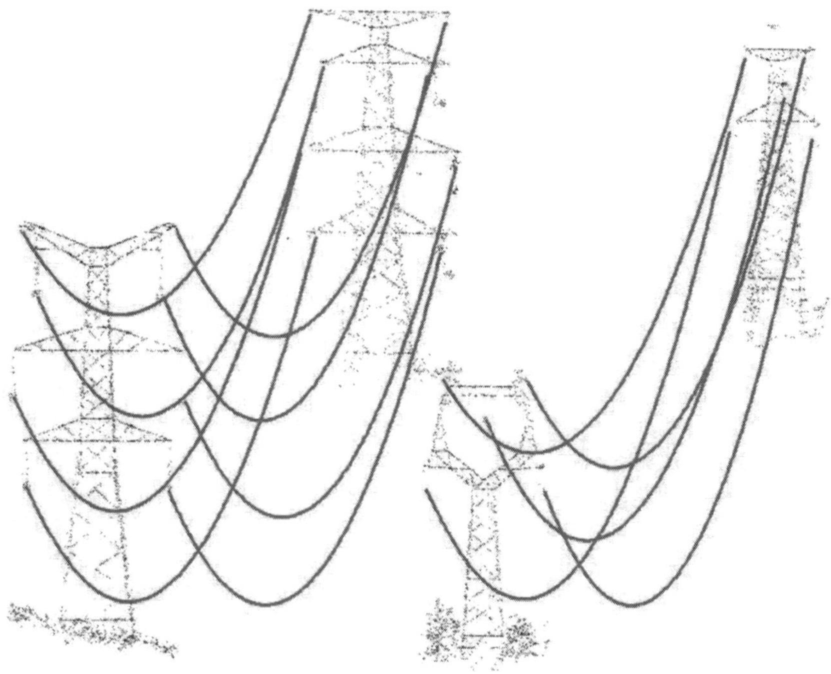

Figure 7. Power-line reconstruction results.

One typical situation is the splitting: a group of power-line spans are split to be two or more groups. In this situation, power-lines are connected asymmetrically between different types of pylons. Irregular power-line span shapes increase the difficulty of the reconstruction. Using our method, based on the local shape properties of a single power-line span and the distribution properties of power-line group between two neighbor pylons, power-line spans can be successfully reconstructed (Figure 8).

Power-line spans are found to be reconstructed with accuracies as given in Table 2. As for reconstruction accuracy calculation, all power-line spans were manually detected to produce a reference map for the power-line span models. Relying on intensive visual inspection, we carefully classify the modeling results into four different categories: correct, splitting, undetected, partially detected. Table 2 shows that our method achieved a nearly 90 percent complete modeling rate with a 10 percent error rate. Compared to McLaughlin's method [12], the reconstruction accuracy is higher. The comparing details are shown in Section 5. Due to the robustness, many reconstruction failures such as splitting and undetected spans can be correctly reconstructed.

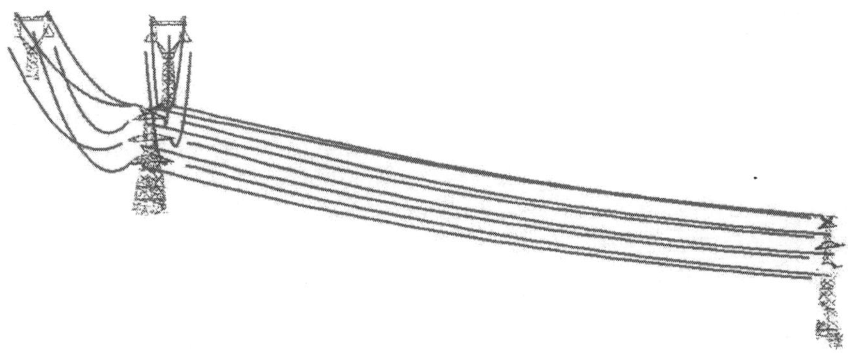

Figure 8. Power-line reconstruction results of splitting situation.

Table 2. Reconstruction accuracy.

–	Correct	Splitting	Undetected	Partially Detected
Our method	89.8%	3.4%	2.7%	4.1%
McLaughlin's method	72.1%	4.6%	12.9%	10.4%

In reconstruction process, we have to analyze the RANSAC iteration parameters and thresholds. Guided sampling in the RANSAC process can substitute random sampling to accelerate convergence and reduces the necessary number of iteration [30]. Based on the similarity detection, the candidate samples on a single power-line span are firstly detected. Thus, the iteration for reconstruct can be reduced. Through tested, the iteration number for reconstructing a span is about 1.2–1.5 times of the number of its data samples.

In seed section detection, neighboring and geometry matching threshold are, respectively, set to be 0.9 and 0.5. As the geometry matching, corresponding points belong to a power-line span match frequently, while the matching rate of wrongly corresponding points is close to zero. Thus, the geometry matching threshold is not large.

During the iteration, samples are added when the residual of fitted function to data is smaller than a threshold of 1.0. The candidate points are then thought to be inliers, or outliers.

Finally, a power-line span model is chosen as the final result if its length is bigger than a defined threshold. The threshold is set according to the distance of two neighboring pylons connected by this span. We set the threshold to be four-fifth of the distance. A piece is considered to be noise if its length is less than one-third of the distance.

Failures of our method are found in some extreme situations. Undetected spans are found where there are almost no detected power-line points (Figure 9). There are two reasons for this failure: (1) It can not to detect an effective seed section of this span using the similarity detection rule if the points are very sparse; (2) The fitting of the span curve function is influenced by the sparseness of points. If the spans within a group are not strictly parallel, the parameters of each span is different. In this situation, the setting of initial parameters for a span with sparse points could not be precise. Therefore, the subsequent fitting of this span may fail.

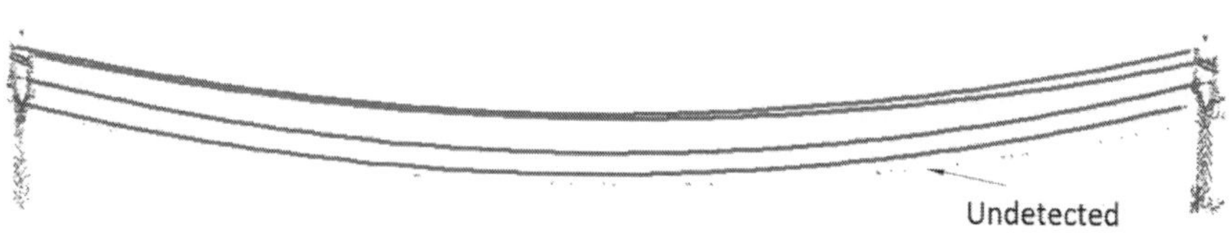

Figure 9. Power-line reconstruction results of undetected spans.

Although the sparseness of points affects the reconstruction results, the quality of reconstruction results are mainly effective for the big gaps caused by sparseness. If the points are regularly distributed, a few number of points can also be well reconstructed. We found that in place with big gap, the span cannot be successfully reconstructed.

Splitting spans are found in places with big gaps, where the length of the gaps are usually bigger than one half of the spans (Figure 10). This failure is also caused by the sparseness of points, resulting a failure to compute a reliable model for these spans. Therefore, the points distributed dispersedly cannot be represented with an unreliable model.

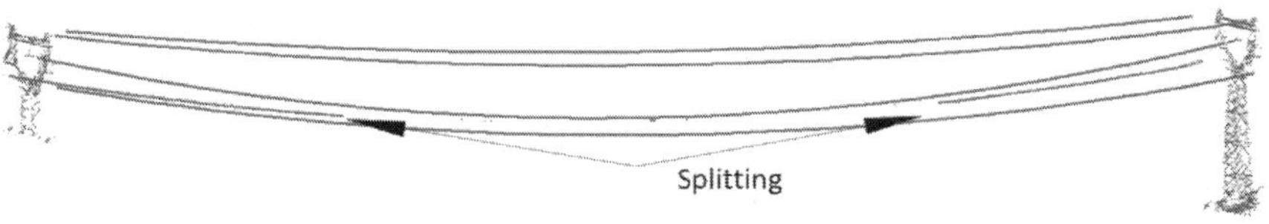

Figure 10. Power-line reconstruction results of splitting spans.

5. DISCUSSION

The ALS technology is often used for automatically monitoring the safety of power-line infrastructure. In this paper, an improved and robust algorithm is proposed to reconstruct power-line spans from ALS data. Before reconstruction, the point cloud data have to be classified and extract the power-line points. Below, we discuss the details of these two stages.

In the initial classification of the data as power-line, vegetation, ground, pylon, and building, the precision and recall of power-line is less than those of other classes. There are many limiting factors in the accuracy of the algorithm. The sparseness of the data is found to be the problematic factor. During the classification process, a local neighborhood is computed for features calculation. The neighborhood we use is with specific shape and fixed size. Some neighborhoods is found to contain few single points. In such situations, data are classified as vegetation by default. Other incorrectly classified neighborhoods are seen to contain an intersection of power-line with vegetation or pylon. Data would be also classified as vegetation. False positives typically occurred with neighborhoods containing very few points of no power-line. The classification accuracy is a key point for power-line reconstruction. With the increasing of classification accuracy, the reconstruction results can be improved.

Due to parts of the power-lines being obscured by vegetation, or a lack of data points along a section of line, there are sparseness and large gaps of the data. Therefore, there may be a small number of power-line spans that are split into several or are undetected. This study was intended to design and test an improved and robust approach for power-line spans reconstruction over the traditional region growing based methods. The additional power-line group distribution and contextual pylons information besides the local shape of spans used in our method are mainly contributed to the improvement.

In order to analyze the robustness, we need to compare our reconstruction results with the traditional methods which only use the local shape properties of a power-line span. The method of McLaughlin [12] is the base of many modified region growing based methods. The main default of this type of methods is similar. In the initial models generation at early growing iteration steps, there might be a very small number of points available due to the effects of occlusions, shadows and system errors. In this case, the estimated model has deviated from the targeted global model [16]. Then, the candidate samples cannot be rightly searched using the region growing rule. Thus, the comparison with the method of McLaughlin is reasonable.

If there are large gaps along one power-line span, using the method of McLaughlin [12], it may result in a splitting failure such that the iterative grouping algorithm can not to aggregate the separated parts into a single span (Figure 11b). Due to robust properties of our methods, these kinds of spans with large gaps can be aggregated into a single span (Figure 11c). Using our method, the initial parameters of a span may be well set. Combining the effective candidates searching of RANSAC, these gaps can be correctly reconstructed.

A number of power-line spans are also undetected using the method of McLaughlin [12] (Figure 12b). This is primarily due to the sparseness of the data, resulting in a failure to fit initial parameters of a local model or a failure of the iterative grouping algorithm to aggregate the sparse points into a single span. Using the methods introduced in this paper, based on the properly-fitted parameters and effective candidate samples searching, this problem can be overcome and produce good results (Figure 12c). In Figure 12c, a span with sparse points is successfully reconstructed except for a small part undetected where there are almost no data points.

Due to the robust properties, the reconstruction results using our method are better than McLaughlin's method [12], which is mainly based on local shape of a power-line span and region growing rule. The improvements are mainly due to the auxiliary information of distribution properties of power-line group set and contextual pylons. The improvements are also based on the RANSAC based reconstruction rule, in which we design reliable initial parameters fitting and efficient candidate sample searching according to the similarity of power-line group set distribution.

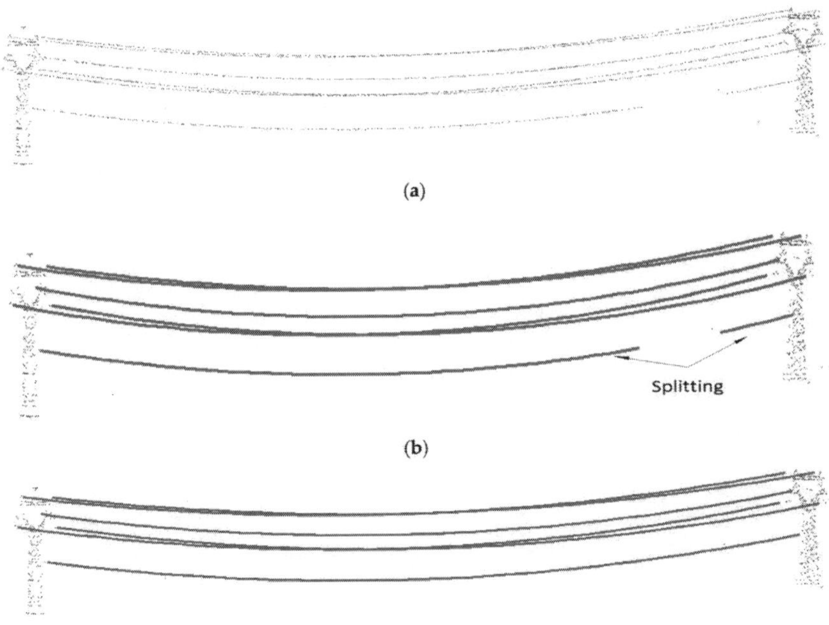

(a)

(b)

(c)

Figure 11. (**a**) Original data; (**b**) reconstruction results using McLaughlin's method [9]; and (**c**) reconstruction results with our method.

6. CONCLUSIONS

In power-line risk management, the robust, precise and automatic modeling of power-lines can provide a great benefit for the execution of a rapid and correct clearance in cases of potential hazards. In this paper, we introduced an improved and robust method for power-line reconstruction. This method can overcome the problems of undetected and splitting of power-line spans that result from only using the local shape information of power-line spans in traditional methods.

The method was mainly based on RANSAC algorithm. Two key parts of initial parameters fitting and candidate samples searching were carefully designed to improve the efficiency and precision of reconstruction. The improvement was also based on using contextual pylon information and distribution property of power-line group. The more information other than local shape information used in reconstruction are mainly described below.

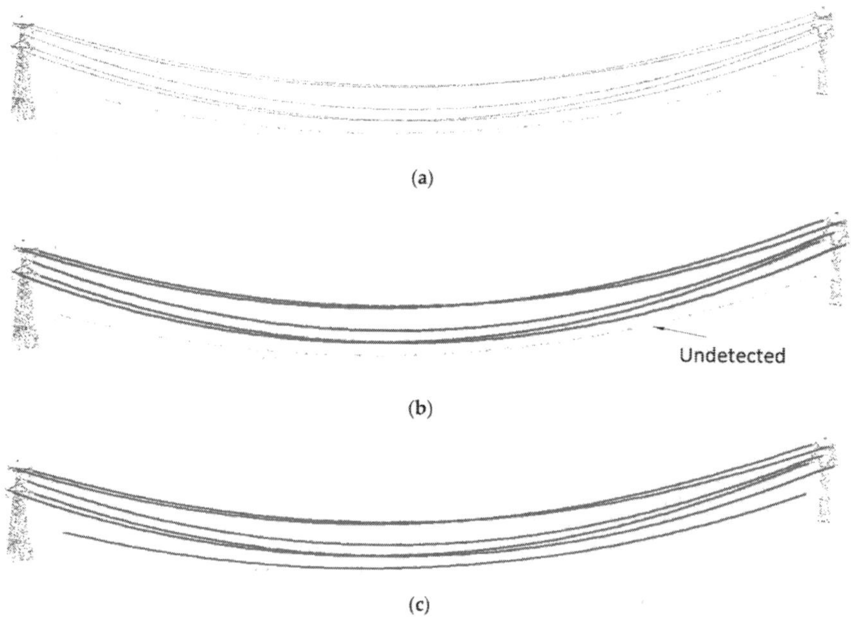

Figure 12. (**a**) Original data; (**b**) reconstruction results using McLaughlin's method [9]; and (**c**) reconstruction results with our method.

Contextual Pylon Information. We automatically classified the point cloud data into five classes: power-line, vegetation, building, ground and pylon. Pylon and power-line are two related kinds of objects. Based on the contextual information, the reconstruction can avoid the defaults caused by only concerning individual power-line object.

Distribution Properties of Power-Line Group. Based on the nearly parallel nature of power-line group between two neighbor pylons, we introduced a similarity detection method to roughly detect the probability of neighbor points belonging to the same power-line span. The distribution properties of power-line group can assist the power-line reconstruction.

Generally, it deviates from traditional methods of object reconstruction, which have been mainly based only on the properties of a single kind of object. Using related and auxiliary information can improve the reconstruction results.

Due to the detection rate of power-line points and data gaps, there are still several defaults of the performance of our reconstruction method. To improve the robustness, the post-processing of power-line models should be carried out by rectifying the incompletion of power-line models based on multi-span level analysis including inner- and across-span in future. Furthermore, the sag of power-line span is changing in time according to the ambient conditions such as the temperature and the ageing of span. Thus, it is necessary to regularly check. In the future, the potential factors that effect the sag (including ambient air temperature, electrical load, solar radiation, wind, etc.) should also be considered to achieve precise modeling of power-lines.

ACKNOWLEDGMENTS

The authors would like to thank the anonymous reviewers for their comments. Their insightful suggestions significantly improved this paper. This work was supported in part by the Major State Basic Research Development Program (973 Program) under Grant No. 2012CB725300, the National Science Foundation of China (Grant No. 41501499 and Grant No. 41371377), Shenzhen Scientific Research and Development Funding Program (No. JCYJ20150625102531697 and No. ZDSY20121019111146499), Postdoctoral Science Foundation of China (Grant No. 2015M572365), and the Open Foundation of Key Laboratory of Precise Engineering and Industry Surveying of National Administration of Surveying, Mapping and Geoinformation (Grant No. PF 2015-15).

AUTHOR CONTRIBUTIONS

Bo Guo, Qingquan Li and Xianfeng Huang contributed equally to the design of this study and the writing of this manuscript including the figures and tables. Chisheng Wang contributed significantly to the discussion of results and manuscript refinement.

CONFLICTS OF INTEREST

The authors declare no conflict of interest.

REFERENCES

1. Keshavarzian, M.; Priebe, C.H. Sag and tension calculations for overhead transmission lines at high temperatures-modified ruling span method. IEEE Trans. Power Deliv. **2000**, 15, 777–783.

2. Mills, S.J.; Castro, M.P.G.; Li, Z.; Cai, J.; Hayward, R.; Mejias, L.; Walker, R.A. Evaluation of aerial remote sensing techniques for vegetation management in power-line corridors. IEEE Trans. Geosci. Remote Sens. **2010**, 48, 3379–3390.

3. Frank, M.; Pan, Z.; Raber, B.; Lenart, C. Vegetation management of utility corridors using high-resolution hyperspectral imaging and lidar. In Proceedings of the 2nd Workshop on Hyperspectral Image and Signal Processing: Evolution in Remote Sensing (WHISPERS), Reykjavik, Iceland, 14–16 June 2010.

4. Hayes, R.M.; Nourai, A. Power Line Sag Monitor. U.S. Patent No. 6205867, 27 March 2001.

5. Turner, R.; Panciera, R.; Tanase, M.A.; Lowell, K.; Hacker, J.M.; Walker, J.P. Estimation of soil surface roughness of agricultural soils using airborne lidar. Remote Sens. Environ. **2014**, 140, 107–117.

6. Niemeyer, J.; Rottensteiner, F.; Soergel, U. Contextual classification of lidar data and building object detection in urban areas. ISPRS J. Photogramm. Remote Sens. **2014**, 87, 152–165.

7. Ashidate, S.-I.; Murashima, S.; Fujii, N. Development of a helicopter-mounted eye-safe laser radar system for distance measurement between power transmission lines and nearby trees. IEEE Trans. Power Deliv. **2002**, 17, 644–648.

8. Motomura, T.; Hirao, M.; Nakamura, K.; Kikuchi, T. Development of the heliborne separation measuring system and its practical application. In Proceedings of the 1998 IEEE 8th International Conference on Transmission and Distribution Construction, Operation and Live-Line Maintenance (ESMO'98), Orlando, FL, USA, 26–30 April 1998; pp. 203–209.

9. Axelsson, P. Processing of laser scanner data—Algorithms and applications. ISPRS J. Photogramm. Remote Sens. **1999**, 54, 138–147.

10. Clode, S.; Rottensteiner, F. Classification of trees and powerlines from medium resolution airborne laserscanner data in urban environments. In

Proceedings of the 2005 APRS Workshop on Digital Image Computing (WDIC), Brisbane, Australia, 21 February 2005; pp. 191–196.

11. Zhu, L.; Hyyppä, J. Fully-automated power line extraction from airborne laser scanning point clouds in forest areas. Remote Sens. **2014**, 6, 11267–11282.

12. McLaughlin, R.A. Extracting transmission lines from airborne lidar data. Geosci. Remote Sens. Lett. **2006**, 3, 222–226.

13. Jwa, Y.; Sohn, G.; Kim, H. Automatic 3D Powerline Reconstruction Using Airborne Lidar Data. Available online: http://www.isprs.org/proceedings/XXXVIII/3-W8/papers/105_laserscanning09.pdf (accessed on 12 October 2015).

14. Melzer, T.; Briese, C. Extraction and modeling of power lines from als point clouds. In Proceedings of the 28th Workshop of the Austrian Association for Pattern Recognition, Hagenberg, Austria, 17–18 June 2004; pp. 47–54.

15. Liang, J.; Zhang, J.; Deng, K.; Liu, Z.; Shi, Q. A new power-line extraction method based on airborne lidar point cloud data. In Proceedings of the 2011 International Symposium on Image and Data Fusion, Tengchong, Yunnan, 9–11 August 2011; pp. 1–4.

16. Jwa, Y.; Sohn, G. A piecewise catenary curve model growing for 3D power line reconstruction. Photogramm. Eng. Remote Sens. **2012**, 78, 1227–1240.

17. Sohn, G.; Jwa, Y.; Kim, H.B. Automatic powerline scene classification and reconstruction using airborne lidar data. Available online: http://www.isprs-ann-photogramm-remote-sens-spatial-inf-sci.net/I-3/167/2012/isprsannals-I-3-167-2012.pdf (accessed on 12 October 2015).

18. Cheng, L.; Tong, L.; Wang, Y.; Li, M. Extraction of urban power lines from vehicle-borne lidar data. Remote Sens. **2014**, 6, 3302–3320.

19. Otcenasova, A.; Hoger, M.; Altus, J. Possible use of airborne LiDAR for monitoring of power lines in slovak republic. In Proceedings of the 2014 IEEE 15th International Scientific Conference on Electric Power Engineering (EPE), Brno, Czech Republic, 12–14 May 2014; pp. 477–481.

20. Lu, M.L.; Kieloch, Z. Accuracy of transmission line modeling based on aerial lidar survey. IEEE Trans. Power Deliv. **2008**, 23, 1655–1663.

21. Fischler, M.A.; Bolles, R.C. Random sample consensus: A paradigm for model fitting with applications to image analysis and automated cartography. Commun. ACM **1981**, 24, 381–395.

22. Guo, B.; Huang, X.; Zhang, F.; Sohn, G. Classification of airborne laser scanning data using jointboost. ISPRS J. Photogramm. Remote Sens. **2015**, 100, 71–83.

23. Bishop, C.M. Pattern Recognition and Machine Learning; Springer New York: New York, NY, USA, 2006; Volume 4.

24. Schapire, R.E. The boosting approach to machine learning: An overview. In Nonlinear Estimation and Classification; Springer: Berlin, Germany, 2003; pp. 149–171.

25. Torralba, A.; Murphy, K.P.; Freeman, W.T. Sharing visual features for multiclass and multiview object detection. IEEE Trans. Pattern Anal. Mach. Intell. **2007**, 19, 854–869.

26. Munoz, D.; Bagnell, J.A.; Vandapel, N.; Hebert, M. Contextual classification with functional max-margin markov networks. In Proceedings of the 2009 IEEE Conference on Computer Vision and Pattern Recognition (CVPR), Miami, FL, USA, 20–25 June 2009; pp. 975–982.

27. Torralba, A.; Murphy, K.P.; Freeman, W.T. Contextual Models for Object Detection Using Boosted Random Fields. Available online: http://machinelearning.wustl.edu/mlpapers/paper_files/NIPS2005_704.pdf (accessed on 12 October 2015).

28. Shotton, J.; Winn, J.; Rother, C.; Criminisi, A. Textonboost for image understanding: Multi-class object recognition and segmentation by jointly modeling texture, layout, and context. Int. J. Comput. Vis. **2009**, 81, 2–23.

29. Zitova, B.; Flusser, J. Image registration methods: A survey. Image Vis. Comput. **2003**, 21, 977–1000.

30. Choi, S.; Kim, T.; Yu, W. Performance evaluation of ransac family. In Proceedings of the 2009 British Machine Vision Conference, London, UK, 7–10 September 2009; pp. 355–361.

31. Wang, Z.; Zhang, L.; Fang, T.; Mathiopoulos, P.T.; Tong, X.; Qu, H.; Xiao, Z.; Li, F.; Chen, D. A multiscale and hierarchical feature extraction method for terrestrial laser scanning point cloud classification. IEEE Trans. Geosci. Remote Sens. **2015**, 53, 2409–2425.

32. Aijazi, A.K.; Checchin, P.; Trassoudaine, L. Segmentation based classification of 3D urban point clouds: A super-voxel based approach with evaluation. Remote Sens. **2013**, 5, 1624–1650.

33. Truong-Hong, L.; Laefer, D.F.; Hinks, T.; Carr, H. Combining an angle criterion with voxelization and the flying voxel method in reconstructing building models from lidar data. Comput. Aided Civ. Infrastruct. Eng. **2013**, 28, 112–129.

34. Kim, H.B.; Sohn, G. Random forests based multiple classifier system for power-line scene classification. In Proceedings of the 2011 Laser Scanning (ISPRS), Calgary, AB, Canada, 29–31 August 2011.

35. Johnson, A.E.; Hebert, M. Using spin images for efficient object recognition in cluttered 3D scenes. IEEE Trans. Pattern Anal. Mach. Intell. **1999**, 21, 433–449.

36. Li, N.; Huang, X.; Zhang, F.; Wang, L. Registration of aerial imagery and lidar data in desert areas using the centroids of bushes as control information. Photogramm. Eng. Remote Sens. **2013**, 79, 743–752.

37. Padgett, C.; Kreutz-Delgado, K.; Udomkesmalee, S. Evaluation of star identification techniques. J. Guid. Control Dyn. **1997**, 20, 259–267.

38. Groth, E.J. A pattern-matching algorithm for two-dimensional coordinate lists. Astron. J. **1986**, 91, 1244–1248.

39. Press, W.H.; Flannery, B.P.; Teukolsky, S.A.; Vetterling, W.T. Numerical Recipes: The Art of Scientific Computing, 3rd ed.; Cambridge University Press: Cambridge, UK, 2007.

CHAPTER 2

The use of airborne laser scanning to develop a pixel-based stratification for a verified carbon offset project

Jordan Golinkoff[1]*, Mark Hanus[2] and Jennifer Carah[3]

[1]The Conservation Fund, 14951 “A” Caspar Road, Box 50, Caspar, CA 95420,USA.

[2]GeoDigital International, McMaster Innovation Park, 175 Longwood Road South, Suite 400A, Hamilton, ON L8P 0A1, Canada.

[3]The NatureConservancy, California Regional Office, 201 Mission St., 4th Floor, San Francisco, CA 94105, USA.

ABSTRACT

Background

The voluntary carbon market is a new and growing market that is increasingly important to consider in managing forestland. Monitoring, reporting, and verifying carbon stocks and fluxes at a project level is the single largest direct cost of a forest carbon offset project. There are now many methods for estimating forest stocks with high accuracy that use both Airborne Laser Scanning (ALS) and high-resolution optical remote sensing data. However, many of these methods are not appropriate for use under existing carbon offset standards and most have not been field tested.

Results

This paper presents a pixel-based forest stratification method that uses both ALS and optical remote sensing data to optimally partition the variability across an ~10,000 ha forest ownership in Mendocino County, CA, USA. This new stratification approach improved the accuracy of the forest inventory, reduced the cost of field-based inventory, and provides a powerful tool for future management planning. This approach also details a method of determining the optimum pixel size to best partition a forest.

Conclusions

The use of ALS and optical remote sensing data can help reduce the cost of field inventory and can help to locate areas that need the most intensive inventory effort. This pixel-based stratification method may provide a cost-effective approach to reducing inventory costs over larger areas when the remote sensing data acquisition costs can be kept low on a per acre basis.

KEYWORDS

Forest carbon offsets, MRV LiDAR, Airborne Laser, Scanning stratification post-stratification, carbon project, carbon stock estimation

1. BACKGROUND

The world's forests are a critical sink of carbon dioxide [1]. It is estimated that forest degradation or destruction results in 6 to 17% of total anthropogenic CO_2 emissions annually [2]. Because of the importance of forest ecosystems in adapting to and mitigating climate change, there are now many policy initiatives to preserve and restore forest ecosystems for a climate benefit [3, 4]. Despite years of discussion however, policies to reduce emissions from terrestrial ecosystems have generally not been adopted. An exception to this is California's cap and trade system that will incorporate carbon offsets starting in 2012 (barring a legal challenge) - see [5].

In part due to the dearth of climate change policies, a vibrant voluntary carbon offset market has sprung up centered around a suite of different carbon project standards [6–9], and managing forests for carbon offsets can provide an important income stream for landowners willing to undertake the costs and requirements of these standards. These standards all have slightly different requirements regarding how to quantify the amount of carbon offsets generated, but generally all require periodic ground-based installation and measurement of plots to monitor project level carbon storage. This paper will focus on the requirements of the Climate Action Reserve Forest Project Protocol as this protocol is substantially similar to what will likely be adopted by the state of California for their compliance carbon market system. The ground based inventory described here, like most traditional forest monitoring, relies on tree measurement and conversion to volume, biomass, and carbon equivalents using established species-specific regressions developed through destructive sampling of trees [10–13]. These sample-based estimates of forest carbon storage are then extrapolated across the full project, often through a stratification approach, whereby unsampled areas receive estimates from areas with similar characteristics based on their remotely sensed attributes [14].

This traditional approach to estimating forest parameters has recently been supplemented and improved upon with the use of remote sensing technologies like Light Detection and Ranging data (LiDAR) paired with high resolution multi-spectral imagery. While these new technologies can accurately estimate forest carbon stocks and fluxes, some of the methods are not easily applicable to forest carbon offset projects because of their complexity and expense. There is a need to apply these new remote sensing products in the context of the voluntary carbon market to show their usefulness at a project level in conformance with typical forest carbon project standards.

1.1 ALS and Optical Remote Sensing

Optical remote sensing products derived from airborne and satellite-borne sensors - Landsat Thematic Mapping Imagery [15, 16], IKONOS imagery [17], Quickbird imagery [17–20], SPOT HRG imagery [21], Moderate Resolution Imaging Spectroradiometer (MODIS) [22–28], and others [29, 30]- have all been used to classify forest landscapes and in some cases to estimate standing carbon stocks. However, estimates of carbon stocks and classifications created using optical sensors alone usually have trouble differentiating areas with high carbon stocks [31, 32]. Synthetic Aperture Radar (SAR) sensors can help improve estimates of biomass but these sensors also saturate in high biomass systems [33]. Because of these limitations, the estimation of forest carbon stocks is often greatly improved with data about forest structure and specifically forest height. Airborne Laser Scanning (ALS), provides a richer summary of forest conditions and more accurate estimates of volume and biomass due to its ability to accurately capture forest heights (LiDAR intensity values can also be used to improve estimates).

ALS paired with other optical remote sensing data is a well-established approach to spatially estimating forest attributes [34–40]. The use of optical remote sensing data in conjunction with LiDAR data is helpful in both delineating crown boundaries and in differentiating between species [32, 35, 37–40]. The ability to make species level distinctions is especially important when estimating merchantable timber volumes and biomass, as these parameters differ between species in trees that are the same size.

ALS data is collected from an instrument that is flown over the forest on an airplane or helicopter. Laser pulses emitted from an airborne instrument reflect off of terrain and vegetation revealing both forest structure (e.g. - height, sub-canopy elements) and a detailed digital elevation model [41, 42]. Individual laser returns can be discrete or continuous (waveform). The spatial resolution can vary from many returns per square meter to sparser returns. The coverage of the ALS can range between full coverage of a given area with no gaps to a sample of the area based on transects below the flight lines to spot samples within transects (i.e. GLAS) [43, 44].

There are two broad categories of ALS data analysis approaches: area based approaches (ABA)/statistical canopy height distribution approaches, and individual tree crown approaches (ITC). Many individual tree approaches use the cloud of LiDAR point data and their relationship to neighbourhood points to build individual crown polygons and/or 3-dimensional tree profiles [42, 45, 46]. These individual tree records can then be aggregated to any scale required to create stand level estimates. These ITC approaches use both parametric and non-parametric approaches [47].

In area based approaches, plot level data is related to remote sensing data that has been aggregated to pixel, plot, or polygon (e.g. stand) units to estimate volume, biomass, or other area based metrics. Area based approaches fall broadly into two main categories:

1) The first category relates grid-cell or stand level remote sensing data to measured plot characteristics to build parametric models to represent forest data. These models have been shown to explain the vast majority of the variation in tree height, diameter at breast height, volume, biomass, basal area, and a suite of other parameters [36–38, 41, 48–51]

2) The second broad category uses non-parametric classification or nearest neighbour methods to stratify the forest into similar groups [52–58]. Non-parametric approaches include k-nearest neighbour techniques [59] and classification algorithms such as Random Forests [53].

Area-based approaches and individual tree approaches to estimating forest parameters are not mutually exclusive however, and several authors have shown how area based systems can be combined with individual tree methods [40, 60].

1.2 ALS and Optical Remote Sensing for a Forest Carbon Offset Project

The methods outlined above all provide different approaches to using ALS data and other data sources to estimate forest parameters. There are two main hurdles in using these methods for forest carbon offset projects. First, the method must be cost-effective and must also fit within the existing management framework of the project. Second, the estimation method must meet the monitoring and verification requirements of the carbon offset protocol. These protocols require periodic inventory of the forest and the application of species-level biomass and carbon conversion equations to all inventory estimates [7, 8, 61]. For example, the Climate Action Reserve Forest Project Protocol v3.2 requires that the United States Forest Service biomass conversions are used for all trees in the project area. Using a stratified inventory approach provides an easily understandable way to generate strata-level tree lists simply from plot data and because of this is more easily verified [8]. Although it may be possible to use some of the existing approaches within a forest carbon project framework, their complexity makes them difficult to understand and potentially challenging to verify. Some approaches do not generate species specific estimates of tree size that can then be used to expand to volume and/or biomass using approved biomass regressions (e.g. - [36]). **The primary objective of this paper will be to describe how the ALS and optical remote sensing stratification system adequately meets the requirements of forest carbon protocols while improving the accuracy of forest inventory estimates.**

In addition to describing a method for ALS and optical remote sensing data to stratify a forest ownership to meet the requirements of a carbon project protocol, this paper will also detail how and where sampling should occur. ALS and optical remote sensing data provide a wealth of information that can be used to increase the efficiency of sampling a forest. **A secondary objective of this paper then, is to provide a method to choose the optimal size for the units of analysis (grid-cell size) and to locate plots**

across the project once the grid is established. Past research has used LiDAR data to stratify an area and locate field plots but these studies have not combined both LiDAR and optical data in the stratification and plot location. These studies have shown that using LiDAR data to first stratify an area and then to locate field plots based on initial strata reduced the root mean squared error (RMSE) of predicted volume [44, 62].

The question of the optimal grid-cell size has been addressed from the opposite direction by Gobakken and Næsset [63]. They examined the optimum plot size to use to best correlate the remote sensing data with the inventory data; however their analysis only used fixed area plot designs and did not examine what scale to aggregate the remote sensing data (i.e. - how big should the grid cells be?). Van Aardt et al. [64] examined various sizes of stands using variable radius plots but their analysis involved the best fit when a stand could contain multiple plots and did not use a regular grid system. Therefore, this new approach will show how to find the most appropriate grid cell size that relates variable radius prism plots to remotely sensed data where each grid cell receives no more than one plot.

Although there has been ample discussion of the technical nature of ALS-assisted forest estimation, few studies move beyond the initial analysis and results with an eye to future management and monitoring. **The third and final objective of this study is to examine how to best leverage data generated by this stratification and modelling exercise for typical management purposes and how to perform inventory updates assuming regular remote sensing data acquisition is not feasible (given cost constraints).**

Using an ALS and optical remote sensing stratification system, a verified and registered carbon project in Mendocino County, California, the Garcia River Forest (GRF), was inventoried in 2010 to meet the requirements of the California Climate Action Reserve (CAR) Forest Project Protocol. Three remotely sensed image datasets - color infrared data (CIR), Red, Green, and Blue true colour imagery (RGB), and LiDAR data - were used to create a canopy segment layer, a canopy height model, and a digital elevation model. These data were summarized to 20 m (1/10 acre) grid cells over the property. An initial systematic random sample was then installed over the full property. The remotely sensed variables were collapsed using a principal components analysis, and combined with the canopy segment summary variables and topographic descriptors, and field survey data to explain the variation in the initial sample of basal area (BA) using a regression model (models to predict trees per hectare (TPH) and percent conifer BA were also developed). The BA model was then used to estimate the basal area for each grid-cell on the property. The BA modelled estimates were then combined with average canopy height derived from the LiDAR canopy height model and the product of basal area and canopy height was calculated as a proxy of volume. This proxy was then divided into classes using an optimal binning heuristic, to define the strata. After this final stratification was completed, a second set of plots were installed to fully inventory each strata, with the number of plots based on the variability of each strata (see Figure 1).

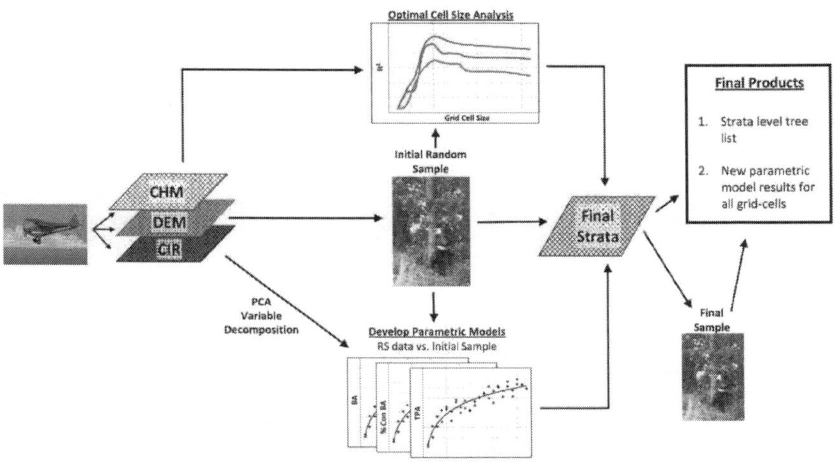

Figure 1. Outline of ALS and optical remote sensing data stratification method.

2. RESULTS

2.1 Traditional Stratification and Inventory and Approaches

Traditional forest stand delineation and stratification (typing) are done by examining aerial photos of a forest and manually drawing boundaries around similar forest areas. This approach requires a forester to then place each stand into a stratum, based on their familiarity with actual forest conditions. This stratification may also use a visual check of ground data and may incorporate some plot data to inform how stands are assigned to strata [65–68].

This approach to stand-delineation and stratification is preferred to unstratified sampling designs, both because of its simplicity and its accuracy in estimating forest parameters. This approach is also preferred because knowing stand boundaries is useful for management purposes and harvest planning. The use of forest strata and stand delineation is ideal in forests with well-documented management histories and/or areas where even-age management was used in the past. Stand boundaries are easily seen and delineated when they correspond to past management and management history can inform the typing of stands. However, in forests managed with uneven-aged silvicultural systems or without a well maintained history of past management, it can be difficult to create a stand map that accurately partitions the variability of a forest due to the relative homogeneity of the forest when observed from aerial photos. In this study, the field site fits within one of these categories: the past management was well-documented but the uneven-aged harvests have left a forest that does not have many clear stand boundaries (see Figure 2), thus rendering the traditional stratification approach less accurate.

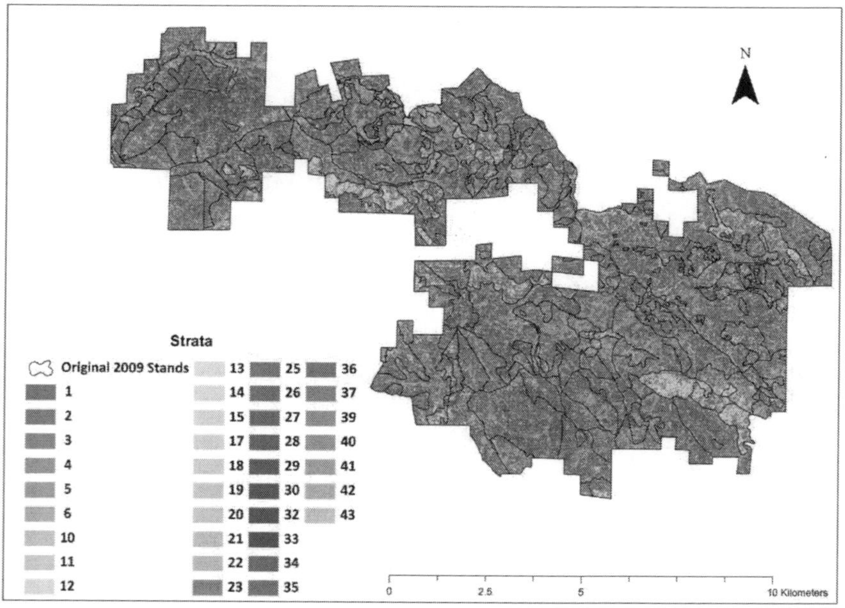

Figure 2. Overlay of 2009 Stand Layer with final stratification of the Garcia River Forest.

Using an ALS and optical remote sensing stratification system, the 9,623 ha (23,780 acre) GRF property was divided into 36 strata (35 forested and 1 non-forested) across the property. Each strata is at least 4.05 ha (10 acres) in size. Strata with higher numbers generally represent better stocked forest areas that have larger trees with more volume and carbon. This approach to forest stratification produces inventory estimates with more statistical confidence relative to the traditionally stand-based inventory approach using about half as many plots (see Table 1 and Table 2). Figure 2 shows a map of the strata generated by this new approach with the old stand boundaries shown in black. Except for the green areas that correspond with grassland, brush-fields, true oak woodlands, or stands treated to reduce tanoak competition most of the property has unclear stand boundaries in a traditional sense, with a high degree of variability within stands.

The original forest accuracy estimates are based on all plots grown forward to 2009 using the Forest Projection and Planning System growth and yield model calibrated to the Northern California redwood region. The 90% accuracy percentage is the property level standard error of the mean multiplied by the 90% t-value (1.645) divided by the mean value.

2.2 Regression Model Results from the Initial 199 Plots

The model form used to explain the correlation in BA is shown below. Both the response and predictor variables have been transformed using a natural logarithm transformation.

$$Y = X\beta + \varepsilon$$

where Y is the transformed response, X is a matrix of transformed predictors identified by the Lasso method and β is the vector of least squares coefficients. The predictor variables used in these regressions are several topographic and LiDAR tree crown variables and the principle components of the color-infrared (CIR) and RGB imagery data sets as well as the PCA rotations for a suite of variables derived from the LiDAR data (the PCA rotations were used to reduce the number of parameters to analyze when building these regressions - see the Appendix for a full list of the predictor variables considered). The components of the β vector and the predictor variables (X) for the BA model are listed in Table 3. The variables are arranged such that those explaining most of the variation are listed first and those explaining the least are last. Regression relationships for trees per hectare and percent conifer BA are also shown below. These relationships were used when lumping strata with less than 10 acres into other larger strata in the last step of the stratification process. A logistic model form was used for % Conifer BA.

Table 1. Inventory Accuracy Statistics

Sample Type	Original Forest Inventory: (Multi-Stage Probability Proportional To Size Stand Based Stratification)	ALS and ORS Grid-Based Inventory: (post-stratification)
C 90% Accuracy	3.72%	3.42%
BA 90% Accuracy	5.4%	3.60%
BF 90% Accuracy	7.56%	5.30%

The original forest accuracy estimates are based on all plots grown forward to 2009 using the Forest Projection and Planning System growth and yield model calibrated to the Northern California redwood region. The 90% accuracy percentage is the property level standard error of the mean multiplied by the 90% t-value (1.645) divided by the mean value.

Table 2. Summary and Comparison of 2009 and 2010 Stratification Systems

	2009	2010
Total Plots	1579	810
Max Plots/Strata	394	40
Min Plots/Strata	4	15
Median Plots/Strata	45	22
Average Plots/Strata	75	23
Total Stands (Pixels)	278	240,410
Sampled Stands (Pixels)	170	810
Max Stand (Pixel) Area (ha)	1,023	0.04
Min Stand (Pixel) Area (ha)	0.8	0.04
Median Stand (Pixel) Area (ha)	14	0.04
Mean Stand (Pixel) Area (ha)	33	0.04
Forested Strata #	21	35
Max Strata (ha)	1,704	1,816
Min Strata (ha)	7.3	3.9
Median Strata (ha)	230	76
Average Strata (ha)	444	255

The 2010 "stands" are called stands as that is their closest analogue when thinking about a traditional stand-based stratified forest inventory. However, these "stands" do not correspond to management units and are therefore better thought of as pixels.

All coefficients are significantly different from 0 at the 95% confidence level.

As has been found in previous crown-based inventory projects, the LiDAR and CIR based variables predict the BA and TPH components best, while LiDAR and RGB variables are more help in predicting species composition [32, 37–40]. The dominance of the color variables in predicting species composition is likely due to the realized species composition of the property being better represented by the image spatial domain than the image frequency domain. The spatial domain treats the image plane as a spatially related database and summarizes the pixel information in context to its neighbors. The frequency domain works on the Fourier transformation of the pixel information. In this case texture, characterized by both grain size and arrangement were more important in discerning species composition than were the absolute color values [69]. In other environments where the leaf color differs more profoundly, color has been more important than texture.

Table 3. Final Model Forms and Coefficients

BA		TPA		% Conifer BA	
Intercept	3.079788313	Intercept	6.19851	Intercept	-0.04949619
CIR3	-0.11917071	Crown closure	0.0006754	LI1	0.161971603
Average crown segment height	0.00519755	LI6	-0.19544	RGB4	0.81924046
Crown closure	0.017182801	LI7	0.05154	LI2	0.09321113
LI7	0.07755464	LI4	0.02984	LI6	-0.19769152
		CIR6	-0.11007	RGB5	-0.50907623
		LI2	-0.20571	LI7	0.294606256
		LI1	0.18478	RGB1	0.824221728
				RGB6	-0.42129326
				LI5	-0.50907623

All coefficients are significantly different from 0 at the 95% confidence level.

Since the coefficient of determination (R^2) is the square of the correlation (ρ) between the predicted and observed values, a simple transformation of it provides a measure of the sample efficiency (Table 4). Sample efficiency is the ratio of the number of correlated plots to uncorrelated plots required to achieve the same level of precision [70]. For example, using values from the table 4 a sample correlated to BA only would require 40.6% of the plots to achieve the same level of significance compared to an uncorrelated sample. This analysis is complicated since the goal of this project is to predict forest structure, which is a composite of these values (and others). The sampling efficiency therefore cannot be directly derived from these values; they are provided strictly as an illustration. However, if forest structure can be reduced to a single metric and that metric related to the remotely sensed data it is likely that the sample reduction would be even more significant (future efforts will likely sample based on Board Foot volume or total volume as this is more related to forest structure). Board Foot (BF) volume is the merchantable volume of trees and only is calculated for merchantable trees (i.e. - conifer species). This value is important for management purposes as BF volume is the primary economic value of many forests.

Table 4. Initial Model Fit Statistics

Model	MSE	R^2	Sample efficiency = 2 (1-ρ)	Number of variables
BA	0.21687	0.635	40.6%	4
TPA	1.46939	0.568	49.3%	7
% ConBA	1.95837	0.493	59.1%	10

Figure 3 shows the modeled versus measured BA in the original and final plots. An examination of the model fit with the original 199 plots (blue) showed that there weren't any strong trends in the residuals.

2.3 Final Stratification Results

The final ALS-optical remote sensing stratification system resulted in more accurate property level estimates of live and dead carbon and basal area than the prior traditional stratification system (Table 1). Accurate stand delineation has the goal of maximizing between-stand variance while minimizing within-stand variance. To better understand the improvement this new approach to stratification provides, it is compared to the previous inventory that used a traditional stand-based stratification.

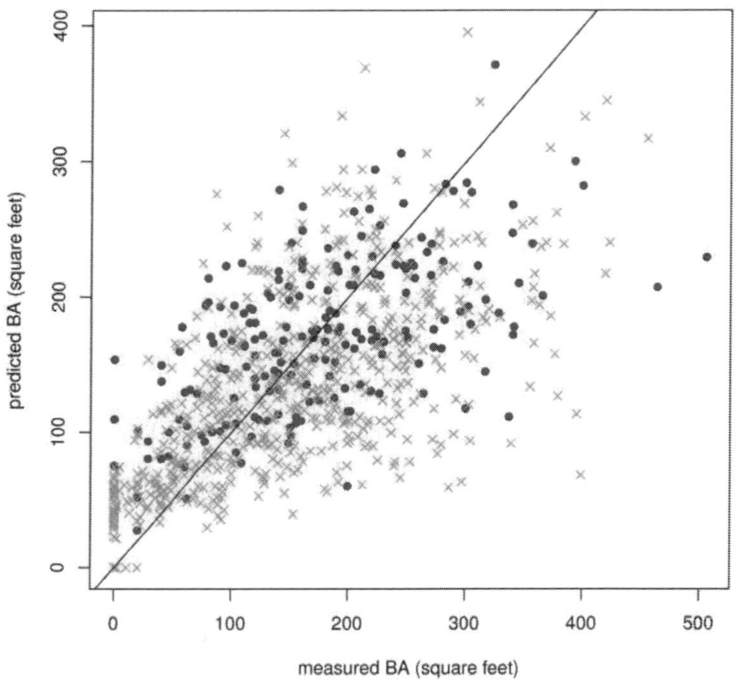

Figure 3. BA Model residuals. Initial sample: blue dots, final sample: orange x's. The BA model residuals were not significantly different than a normal distribution (Pearson Chi-Square Normality Test, p-value = 0.7076)

Based on the results seen using this new stratification approach there are several conclusions that can be drawn. First, with half as many plots (Table 2), we have more statistical confidence in the inventory using this method due to the high resolution stratification derived from the remotely sensed imagery (Table 1). Second, this new stratification approach has shown that past samples most likely averaged more highly stocked riparian areas with non-riparian areas and therefore showed less volume on this property. Third, this new strata system allows for a flexible approach that can be easily leveraged when designing timber harvest plans or trying to understand the habitat of a given area. For example, accurate inventory estimates can now be made for any polygon across the full ownership simply by aggregating a set of grid cells.

2.4 Old Stand Level Comparison

Visually, the strata systems are much different (see Figure 2 and Figure 4), as the old stand boundaries lump together many cells that are currently considered different strata. This visual comparison shows that although the old stratification and stand delineation does a reasonable job of capturing some of the differences in the stands, there are many areas where it is hard to see well defined stand boundaries.

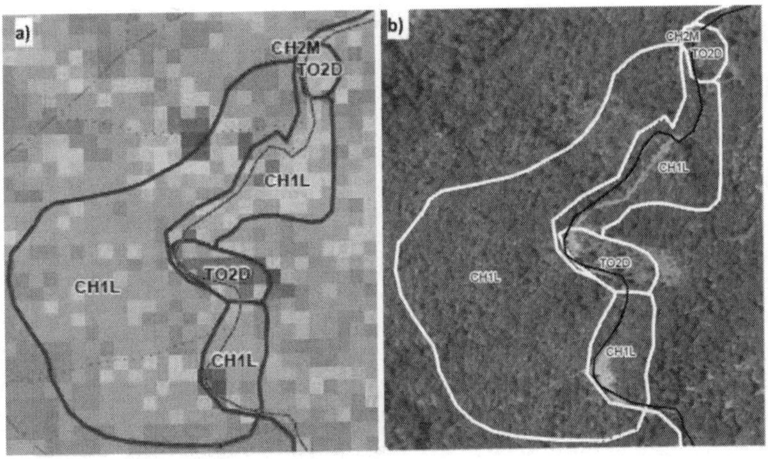

Figure 4. A visual comparison of the current stratification system versus the prior system. Figure 4a shows the current strataification system (with the old stand boundaries as well). The 0.04 ha grid cells are shaded to represent their different strata with redder cells having less volume than green cells. Figure 4b shows the prior stand delineation for this same area with the true-color imagery of the area as the base layer to show actual forest conditions. Note that the new strata grid-cells do correspond to the old stratification in areas where there are clear stand boundaries but in mixed forest conditions the new system can distinguish different forest conditions that the original strata system lumped together. This new strata system also does a much better job of mapping landings/clearings and wide road areas (most of the red and orange cells).

Another way to compare the current strata system to the prior system is to look at some well sampled stands in the prior inventory and compare those estimates to the current strata-based estimates (Table 5). Quantitatively the differences between estimates of stand parameters are not statistically significant (except for BA - this result was also found in Hudak et al. [52] and they postulate that this bias is a result of the natural logarithm transformations and back transformations). These results therefore are an indication that the current stratification system, though much different than the previous system, produces estimates of stand level parameters that are similar to a traditional forest inventory (but more accurate). The advantage is that these estimates can now be found for any arbitrary polygon across the forest by grouping cells of interest and generating estimates for this group [52]. This approach therefore presents a much more flexible set of data to gauge forest conditions.

3. DISCUSSION

3.1 Selection of Grid Size

The first step in partitioning the variability of the GRF was to establish a grid across the whole property. Many LiDAR driven forest inventories in past studies have used stem-mapped plots to correlate ground data with remote sensing data by using the actual location of trees and their crowns to build models that relate to the remotely sensed crown polygons and crown heights [37]. In this application however, variable radius plots were used to correlate the vegetation and the cell variability recognized by the LiDAR imagery. Stem mapping was not chosen because it would have been prohibitively expensive due to the high number of stems per ha and the steep terrain. However, because variable radius plots were used it is difficult to know the optimal size for grid-cells given that the size of the plots is variable [65].

The exercise of choosing the size of the grid cells is dependent on several factors. The first consideration is the ability to accurately locate sample plots using handheld GPS units. The GPS units used by the inventory cruisers have accuracies that exceed 10 m (33 feet) 95% of the time [71]. The second factor when choosing the grid size is finding the optimal cell size to reduce the variability between the remote sensing data and the measured plot data. Past studies have shown that it is important to choose a grid size that best matches the size of the plots installed [41, 48]. van Aardt et al. [64] also explored this question using an object based approach (as opposed to pixels, objects are non-uniform areas of similar characteristics) and found only a small loss of accuracy with increasing object size. Pesonen et al. [72] have also examined the optimal fixed area grid cell size but for that study focused on finding the optimum grid cell size when estimating coarse woody debris as opposed to standing trees.

Table 5. Comparison of recently cruised stands using old strata system and current strata system

2009 Data (2008 Plot Data Is Grown to 2009)									2010 Data				
Strata	Stand	Ha	Year Cruised	Plots	BA (m^2/ha)	TPH (> 5 cm)	BF per ha	C (Mg/ha - no dead)	# of 2010 Strata	BA (m^2/ha)	TPH (> 5 cm)	BF per ha	C (Mg/ha - no dead)
DR1M	**2**	53	2009	4	47.3	739.8	35,031	174.5	26	45.7	824.8	28,938	157.6
GX2D	**115**	7	2009	4	25.4	339.7	6,169	123.6	16	38.2	709.6	17,104	132.8
MH2D	**171**	35	2008	4	32.7	1,255.6	32,564	127.8	23	44.0	822.0	24,558	150.9
DR2D	**239**	13	2008	4	19.0	219.0	26,084	93.7	23	42.0	695.5	28,769	145.5
DR3D	**265**	54	2008	4	43.2	883.1	55,819	217.8	29	44.6	737.9	32,592	154.5
CH2M	**269**	183	2008	20	43.1	1,404.7	35,222	156.2	30	48.5	839.4	34,417	169.5
CH2M	**270**	138	2008	16	47.5	1,646.9	28,088	170.3	30	48.3	842.1	34,136	168.2
CH2M	**271**	131	2008	16	40.0	1,745.1	22,248	140.2	29	46.7	808.5	32,671	163.1

The estimates of stand level TPH, Board Feet (BF)/ha, and Metric Tons of Carbon (MgC)/ha showed no statistically significant difference between the past stand delineation estimate and aggregating the current stratification system to the old stand boundaries except for basal area (paired t-test p-values: BA = 0.034, TPH = 0.23, BF/ha = 0.81, metric tons Carbon/ha = 0.7).

Approaching the question of the optimum size to best relate plot data to remote sensing data, Gobokken and Næsset[63] used a Monte Carlo analysis to explore the optimal size of fixed area plots in developing accurate forest inventory estimates. This analysis is similar to our current question but may be difficult to implement in practice as the plots may already be measured or it may not be appropriate to change the plot design mid-sample.

In this case, a 4.6 m^2/ha (20 ft^2/acre) basal area factor (BAF) prism was used on each plot. Generally, a 4.6 BAF prism samples about 0.04 ha but this will change depending on the size of the trees. To test this, the average of the limiting distances of each tree measured in all of the variable radius plots was calculated and the median plot size based on this analysis was determined to be 0.036 ha. However, larger trees would likely be outside of grid cells that are 0.4 ha or smaller. In addition, there is a greater chance that the location of the plot in the field would fall outside of the target grid cell due to the variability in the estimates of location made by the handheld GPS units. Therefore, grid cells less than 0.4 ha (1/10th acre) were deemed too small.

As the grid cell size increases to sizes larger than 0.4 ha, the variability of the forest within the cell (and hence the remote sensing data) increases. Because of this, it was hypothesized that any model that relates plot metrics to summarized grid cell remote sensing data will theoretically perform worse as the size of the cell increases to sizes larger than the plot. For these reasons, a 0.04 ha cell size was used as it was deemed to be the smallest cell size that would contain a 4.6 BAF plot and the location error associated with the handheld GPS units, and result in minimal within cell variability.

After further analysis following the completion of the inventory, the 0.04 ha grid cell size may have been slightly too small to create the strongest relationship between plot values (e.g. - BA, TPH, volume, carbon, etc), topographical data (elevation, slope, aspect), and remotely sensed data (e.g. - orthophoto band intensity). The optimal grid cell analysis was undertaken after the inventory was completed as a means to assess if the pixel size used was the best size and to inform future projects. The approach outlined below is one method that could be used to decide on the size of pixels to divide a forested area into and would ideally be used prior to the final sample. To determine the optimal grid cell size, a sample of the remotely sensed data was taken at each field plot point with a series of increasing circular areas (see Figure 5a). The mean and standard deviation of all remotely sensed variables for each circular region for each data set was then calculated for each size circle. Once the remote sensing derived data had been summarized to each sample size, an exhaustive model selection routine was run to find the best model assuming the best model was defined using Bayes Information Criteria (BIC) [73, 74]. The BIC was used as the metric of model performance because it does not assume that a relationship between explanatory and predictor values exists and has a larger penalty with larger data sets [75]. Once the model with the lowest BIC was chosen for each circular area the amount of variation explained was graphed relative to each other sample size (Figure 5b). In this way, an objective approach to model selection can provide a metric to judge which size grid-cell is optimal. Based on the results seen here, it seems the optimal cell size was about 0.08 ha (1/5 acre). This would be slightly larger than the cell size actually used.

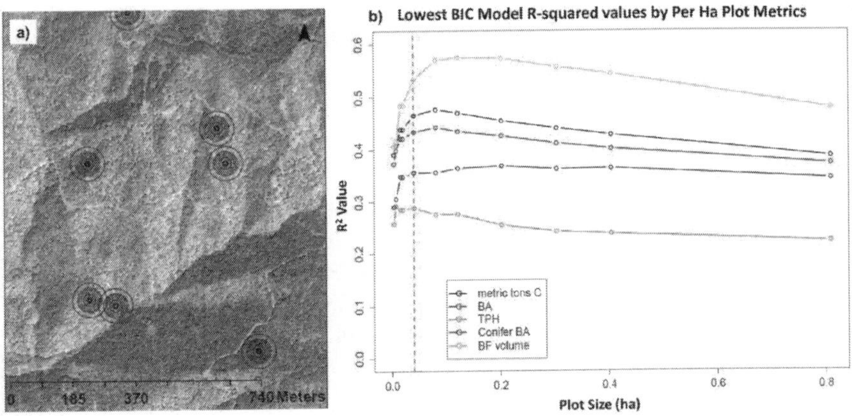

Figure 5 .Optimum Grid Cell Size Results. a) Remote sensing sample units of different size. Red circle represents 0.04 ha. b) Results of lowest BIC model selection approach using an exhaustive search of all potential model permutations. Dashed red line shows 0.04 ha size.

3.2 Sampling Intensity by Strata and Plot Location within Strata

The optimal sampling intensity of the final sample can be determined using a Neyman allocation of plots (or an optimal allocation of plots if the plots have variable costs in different strata) using the traditional approach to estimating the appropriate sample size [66]. For strata that do not have an adequate initial sample to have confidence in the estimate of the sampling variability, an estimate of the variability of the strata can be found using the remote sensing data for that strata compared to the other strata. In this case, using the models developed from the initial strata to populate the cells of the under-sampled strata an estimate of the population variance can be found and used to calculate the optimum sample size. Plots are then randomly located within the strata.

4. FUTURE DIRECTIONS

4.1 Management Planning

Using this new approach will be a significant departure from how forest planning traditionally proceeds using a stand based approach. Using a grid-based stratification, analysis of given forest areas in these small units can provide more fine-grained information about any given area. For example, when laying out timber harvest plan boundaries, these forest strata can be used to more accurately understand current stocking and forest conditions and allow for better layout of plan boundaries and a better description of pre-harvest conditions and habitat.

Although this stratification approach provides much higher resolution data in terms of understanding current forest conditions, there are several challenges to using this approach. To begin, this grid system does not lend itself to easy modeling of future management because the stand structure (400 m^2 pixels) are not logical management units. Secondly, although we have more confidence in the total volume of any given cell across the property, there may be more variation in the species composition within a strata type. This is a result of the fact that total volume, not merchantable volume, was the variable whose variation was optimized during the creation of strata. In future efforts, both total volume and merchantable volume should be considered when creating strata boundaries.

4.2 Sampling of Harvest or Disturbance

As mentioned above, this strata system provides a highly flexible and accurate picture of current forest conditions. Moving forward, as areas are harvested or undergo natural disturbance however, sampling will revert back to a more traditional harvest area (stand) based approach. The reason for this is twofold. First, the cost of collecting new remote sensing data annually prevents the collection of the necessary data to drive this stratification process. Second, the known THP boundaries or disturbance events can be used to generate more accurate stand boundaries. Therefore, future sampling will proceed by first delineating the disturbed area and then sampling within this area to estimate the standing forest stocks post disturbance.

4.3 Ecological Monitoring

We anticipate that the canopy height model will be used in the future to generate a revised Northern Spotted Owl (NSO) habitat model to assist in management of the NSO. One of the benefits of this small grid system is that the final plot data can also be used to develop full parametric models for any variable of interest. In some cases (e.g. canopy cover), models are not required as the variable in question is measured directly by the LiDAR data. In this case, the canopy cover found in trees greater than 28 cm (11in) DBH will be modeled to inform the classification of NSO habitat [76] (traditionally this classification was based on lower resolution ocular estimates).

4.4 Pre-Aggregation for Process Modeling

Hawbaker et al. [62] show that there is a need for ALS to be leveraged across larger landscapes and that ALS can help to create more accurate estimates of biophysical variables at a landscape scale by helping to better define the sampling design used. The method of sampling and stratification outlined in the following section can also be used to both validate process models and to serve as a pre-aggregation framework across a large landscape. Although this method uses ALS and optical remote sensing data with continuous coverage across the landscape it could also be applied to larger scales using a variety of data sources with or without full coverage. Specifically, by running models based on a small set of strata instead of in each grid-cell across a region much more efficient and rapid estimates of ecosystem state can be generated.

Lefsky et al. [32] have shown the value of using ALS combined with Landsat data to construct independent estimates of landscape net primary productivity and net ecosystem productivity to compare with light-use efficiency models or biogeochemistry models. Their work used remote sensing data collected over time to detect change. The strata system developed here will serve as the basis for future biogeochemistry model runs that will also attempt to better estimate ecosystem carbon fluxes at the GRF.

5. CONCLUSIONS

The method described below not only provides a cost effective and flexible approach to stratifying a forest but also has been designed and applied in the context of the requirements of existing forest carbon project protocols. This is highly valuable given that monitoring, reporting, and verifying carbon stocks and fluxes at a project level is the single largest external cost of a forest carbon offset project. Although currently the use of LiDAR approaches for smaller scales still is not cost effective, using a method like this one at scales larger than 10,000 ha (25,000 acres) may pay for themselves by reducing the cost of the field inventory required.

Additionally, the use of both parametric approaches (to develop models from the initial sample) and non-parametric approaches (to partition the variables of interest into strata) provides more power to determine the optimum sampling intensity and location across a large ownership. Furthermore, the 2 stage sample allows for the optimum grid cell size to be found.

For management decisions, this ALS and optical remote sensing stratification design and high-resolution grid allows for more accurate estimates of volume at any scale larger than a 0.04 ha grid cell (1/10 acre). This new strata layer and the data associated with it will serve as a baseline of forest conditions against which future management at the Garcia River Forest can be compared and assessed. Additionally, because of the flexibility built into this method, it can be scaled to much larger or smaller spatial extents. This is valuable for planning both local and larger scale ongoing management and monitoring activities.

6. METHODS

6.1 Study Site

The Garcia River Forest (GRF) project is a 9,623 ha (23,780 acre) forest located in Mendocino County, California northwest of the town of Boonville. This forest is owned by The Conservation Fund (TCF) and is protected by a conservation easement held by the Nature Conservancy (TNC). The goals of the project are to conserve and restore highly productive and biologically diverse forests and streams, and to implement sustainable forest management practices that support the local economy [77]. This region is historically dominated by a mix of redwood (Sequoia sempivirens)and Douglas-fir (Pseudotsuga menziesii) trees but due to decades of industrial timber management and intensive harvesting of this forest there is now a higher than natural amount of Tanoak (Lithocarpus densiflorus) in traditionally conifer dominated stands.

Due to the past management of the GRF, most stands have a mix of young 2nd or 3rd growth redwood and Douglas-fir trees with high proportions of tanoak. Most areas are heterogeneous within stand boundaries and these conditions are the norm across the full ownership. Past management consisted mostly of "thinning from above" - removing the larger, better trees from most stands - and as a result most stands are made up of small, young trees.

Because of the state of the forest today, it is difficult to use a traditional stand mapping approach to delineate areas that are substantially similar. The result of applying the traditional air photo interpretation approach to stand mapping in this forest resulted in the creation of large stands that have high degrees of within stand variability and don't always relate to logical management units (see Figure 2).

6.2 Field Data

6.2.1 2009 Data (used for comparison to 2010 stratification results)

The existing inventory consisted of plots installed over several years using several different cruising protocols. Both variable radius plots and fixed area plots were installed across the property from 1999 to 2009. Most recently (2006 to 2008), all cruising occurred on a 400 by 400 meter (20 by 20 chain) grid that covered the full ownership using 4.6 Basal Area Factor prisms (Table 2). The complete inventory from 1999 to 2009 was grown forward to 2009 using the Forest Projection and Planning System growth and yield model to compare property level estimates in 2009 to the new stratification method in 2010. However, only plot data from 2008 and 2009 was used to compare individual stand level estimates to aggregated pixel estimates (see table 5).

The old stand layer was a traditional timber stand typing done by head's up digitizing stand boundaries using color imagery (acquired in 2004) of the forest. Each stand was then placed within a strata that described the dominant tree size and species based on the professional judgment of the land manager. The old strata types had 3 fields: a 2 digit species code that described the dominant species or species mix, a 1 digit size-class code that described the dominant tree size, and a 1 digit canopy density code that described the degree of canopy closure.

6.2.2 2010 Data (used for stratification)

The 2010 inventory data was collected between June and September of 2010. It consists of 810 variable radius plots that use a 4.6 m^2/ha (20 ft^2/acre) basal area factor (BAF) prism to measure trees at least 14 cm (5.5 inches) DBH. All plots have height measured on all trees (both live and dead) that are tallied in the variable radius plot. In addition to the trees measured in the prism plot, there is a 0.04 ha (1/10th acre) circular plot for understory vegetation, a 0.004 ha (1/100th acre) plot to measure regeneration (trees less than 14 cm DBH), and a 30.5 m (100 ft) transect to measure down woody debris. Table 4 summarizes the current inventory data and the past inventory data. The past 2009 inventory and stand layer was used as a baseline against which to compare the new 2010 ALS based stratification and inventory system.

The field sampled plots for the preliminary sample (199 plots) were a random selection of a 400 m by 400 m (20 by 20 chain) grid. Table 6 lists the summary statistics for the preliminary sample.

Table.6 Initial 199 Plot Summary Statistics

Variable	Min	Mean	Max
BA (m^2/ha)	0	40.73	116.1
TPH (Trees Per ha)	2	2,339	14,944
% conifer BA	0	56.6	100
Average height (m)	7	29	62

6.3 Remote Sensing Data

Both color-infrared imagery and LiDAR data were collected for the full property (Table 7). The color-infrared imagery has 0.6 meter (2 foot) resolution with horizontal accuracy less than 1 meter. The raw LiDAR returns range from 2.5 to 27 returns per square meter with at least 5 returns per square meter for forested areas. The LiDAR data exceeds 15 cm of vertical accuracy and 50 cm of horizontal accuracy. The LiDAR returns were summarized to make a 1 square meter digital elevation map and a 0.5 square meter canopy height model. The CHM is gridded to 0.5 m^2 and based on the interpolated "highest" return within each pixel. In addition to these grids, the LiDAR data were used to generated a crown polygon layer for the full GRF. The crown polygon layer was created using a watershed transformation algorithm applied to the CHM that segmented individual tree crowns that are isolated in height from adjacent regions.

Table.7 Summary of Remote Sensing Data Collected in 2009

	Color Infrared	Light Detection And Ranging
Acronym	CIR	LiDAR
Date Collected	7/1/2009	
Source	Fixed-wing aircraft	
Instrument	Digital Mapping Camera from Zeiss/ Intergraph Imaging	ALTM Gemini from Optech Incorporated
Scale	Full ownership	
Projection	North American Datum 1983 UTM zone 10N	
Resolution	0.6 meter	5 returns/square meter, 24° field of view, 0.44 postings/square meter.
Spectrum	visible and near-infrared (380 nm to 2500 nm)	near-infrared (760 nm to 2500 nm)
Accuracy	Horizontal accuracy sub 1 meter	Horizontal accuracy sub 50 cm Vertical accuracy sub 15 cm
Data Form	4 bands: red, blue, green, and near-infrared	Discrete Waveform with classified returns (ground, mid-canopy, upper-canopy)
Products	Ortho-rectified 4 band CIR	All and first return LiDAR (raw data) 1 m^2 Digital Elevation Model (DEM) 0.5 m^2 Canopy Height Model (CHM)Crown Polygon Layer

7. DESCRIPTION OF THE METHOD

7.1 Data Summarization to 400 m^2 pixels

The first step before any analysis, inventory, or stratification could occur was to summarize all of the remote sensing data to the 400 m^2 grid cells. This involved finding the average and variance of all of the remote sensing data sets (e.g. CIR, RGB, canopy height, crown polygons, topography variables - slope, aspect, elevation, and a whole suite of other variables derived from the remote sensing data in both the spatial and frequency domains). The complete set of variables used for the analysis and a brief description of them are listed in the appendix.

The source data for the cell summaries used in the stratification come from two passive image datasets and summarized LiDAR. The three image sets (CIR, RGB and CHM) were processed with MATLAB's image processing toolbox [78]. The image processing routines work in two domains; the spatial, and the frequency [79]. The pixels from the image data sets are about 0.6 meters on a side. The CHM is treated as a gray scale image where height above the ground is scaled to the gray scale.

7.2 Initial Plot Installation

To develop the final stratification, a set of "training" field plots were installed to find the relationships between plot data and the cell data (e.g. volume, carbon, basal area). To do this, an initial set of 199 plots were installed across the GRF. A random sample of points located at the intersections of a 400 m by 400 m (20 by 20 chain) grid was chosen to cover a broad spatial area.

7.3 Variable Reduction using Principle Component Analysis (PCA)

The 400 m^2 cell data was summarized using principle component analysis to reduce the number of variables. Factor analysis was used to determine how many of the principle components should be retained [80]. Table 8 lists the amount of variation explained by the first eight and the next eight principal components in the each of the image datasets. Based upon the reduction in explained variance and the need to keep the preliminary sample small, the first eight component vectors were selected to represent the data sets in the preliminary sample.

Table. 8 Principle Component Decomposition of the Imagery Datasets

Image set	Variance explained by first eight	Variance explained next eight
RGB	76.00%	13.40%
CIR	75.10%	13.70%
CHM	72.60%	14.70%

The original optical data consisted of 4 bands of data: blue, red, green, and NIR reflectance values. Although it would be possible to analyze this data by combining all 4 bands into one image, instead this optical data was used to create two images: a color-infrared (CIR) image and a Red-Green-Blue (RGB) image. The CIR image combines the red, green, and NIR values. There are two reasons why the red and green bands were included in both the CIR and RGB datasets: 1) to check that the atmospheric correction was applied correctly and 2) to have finer control of the linear combination of the data when conducting the analysis.

Since two of the color bands (red and green) are present in both the CIR and RGB image data, a correlation analysis was conducted to determine the amount of overlap between the principal components of the two datasets. The Pearson correlations with p-values less than 0.05 have an asterisk in Table 9.

Table. 9 Correlation Analysis between the CIR and RGB Principle Component Datasets

PrinComp	RGB1	RGB2	RGB3	RGB4	RGB5	RGB6	RGB7	RGB8
CIR1	0.925*	-0.128	-0.418*	0.137	-0.144*	-0.282*	-0.011	-0.08
CIR2	-0.072	0.977*	0.091	-0.067	0.064	-0.237*	-0.034	-0.048
CIR3	-0.09	-0.243*	0.158*	0.736*	-0.1	0.029	-0.252*	0.068
CIR4	-0.273*	0.068	0.891*	-0.185*	-0.189*	0.382*	-0.328*	-0.018
CIR5	-0.249*	0.101	0.371*	-0.254*	0.931*	0.123*	-0.11	0.0002
CIR6	-0.226*	-0.201*	-0.079	-0.237*	-0.206*	0.840*	-0.056	0.087
CIR7	0.052*	-0.019	-0.378*	-0.041	-0.009	-0.319*	0.890*	-0.084
CIR8	-0.170*	-0.008	0.066	-0.051	0.06	-0.177*	0.337*	0.880*

A quick scan of Table 9 shows that, as expected, some of the principal components are highly correlated. This correlation reduces the efficiency of variable screening methods applied to this data, meaning that more plots will be required to achieve the same level of certainty. The impact of the correlations was examined by repeating the parameterization of the models described below with both data sets separately and then both together.

7.4 Parameterization of Models to Relate Remote Sensing Data to Initial Inventory

The data collected in the first 199 plots was then correlated to the reduced set of remotely sensed variables found using the PCA. Several models were built that related remotely sensed data to the measured plot data in each sampled 400 m^2 cell. However, only the BA regression model, multiplied by each cell's average canopy height, was used by the Tabu Search Algorithm to develop the initial strata. The BA model was then used to predict the BA in all of the 240,410 400 m^2 cells across the full ownership. The .5 m^2 resolution Canopy Height Model (CHM) was averaged across each 20 by 20 m pixel and used to estimate the average canopy height in each pixel (no model was required as this is directly measured by the LiDAR data).

Stepwise procedures have been found to produce poor variable screens [81]. This is partially due to the repeated comparisons not representing the proper elimination probabilities [82]. However there are other problems with the method such as the parameter estimates being biased high, and the standard error of the estimates being too low. This results in F and chi-squared statistics not having the desired distributions [83]. Based upon this the Lasso method [84] was used for the variable screening of the predictive models. The Lasso is a penalized least squares method which selects a set of regression coefficients (β^{Lasso}) as the coefficients that minimize the following equation:

$$\hat{\beta}^{Lasso} = argmin_{\beta}\left\{\sum_{i=1}^{n}\left(y_i - \beta_0 - \sum_{j=1}^{p} x_{ij}\beta_j\right)^2 + \lambda\sum_{i=1}^{p}|\beta_i|\right\}$$

In the above equation, y is an n-length vector of the response variables; X is an n by p matrix of predictor variables. β_0 and β_j are the standard regression intercept and coefficient vectors while the last term is a penalty term applied to each coefficient - lambda is the penalty multiplier that is applied to each estimated coefficient.

To ensure that no single predictor swamps the effects of others, the matrix of predictors (X) is centered and scaled, and then λ is chosen by cross-validation. This means that a portion of the plots are held back from the regression and these plots are then predicted by the resulting regression. The Lambda value is iteratively adjusted to produce the lowest prediction error of this cross-validation. The Lasso serves as a variable selection methodology by selecting few predictors thus alleviating problems attendant to having many potential predictors compared to the number of observations. Furthermore, since the Lasso tends to select only a few of a set of correlated predictors, it also helps reduce problems with spatial correlation [84].

7.5 Final Stratification Using Supervised Classification

Based on the predictions of the BA model described above, an optimal binning process [85, 86] was used to create bins (strata) for each cell based on the product of A and height. The stratum for each cell was determined by minimizing the amount of variation of the product of BA and height in each strata. The product of BA and height is highly correlated to volume and therefore cells within a given strata have similar volume totals. This classification method is considered supervised since it is driven by the initial inventory data collected across the GRF.

Once the supervised classification was completed, to prevent any strata from being less than 4.05 ha (10 acres) in size, an algorithm was applied to swap grid cells that were on the "edge" of each strata into neighboring strata (considering the nearness according to BA, height, Trees Per Hectare (TPH), and % conifer BA). The goal of this algorithm was to minimize the variation covered within a given strata while reducing the total number of strata.

7.6 Selection of Remaining 611 Sample Plots Based on Final Stratification

The final 611 plots were randomly placed within each final stratum in proportion to the variability in product of BA and height. This sampling design is a classic post-stratification design and therefore uses stratified random sampling estimators [66, 87].

ACKNOWLEDGEMENTS

The Garica River Forest is a project of The Conservation Fund, in partnership with the Nature Conservancy, State Coastal Conservancy and the Wildlife Conservation Board. Funding for the remote sensing data acquisition and processing was provided by The Nature Conservancy's Priscilla Bullitt Collins Northwest Forest Fund, The Nature Conservancy, and The Conservation Fund. The authors would like to thank Evan Smith, Scott Kelly, Madison Thompson, and Holly Newberger for their logistical and technical support of this work as well as their review. We'd also like to thank Mark Reynolds, and Katie Andrews for their support of the data collection as well as Steve Running and for his review. Finally, three anonymous reviewers are thanked for their comments which notably improved the quality of the original manuscript.

COMPETING INTERESTS

'The authors declare that they have no competing interests.

AUTHORS' CONTRIBUTIONS

JG: final write-up, conceptual development, analysis of results, description of future uses, development of optimal grid-cell size analysis. MH: processing of raw data, development of method and application to data, write up of method section, critical review and analysis of results. JC: conceptual development, review, description of future uses.

All authors read and approved the final manuscript.

REFERENCES

1. Denman KL, Brasseur G, Chidthaisong A, Ciais P, Cox PM, Dickinson RE, Hauglustaine D, Heinze C, Holland E, Jacobs DM, Lohmann U, Ramachandran S, da Silva Dias PL, Wofsy SC, Zhang X: Couplings Between Changes in the Climate System and Biogeochemistry. In

Climate Change 2007: The Physical Science Basis. Contribution of Working Group I to the Fourth Assessment Report of the Intergovernmental Panel on Climate Change. Edited by: Solomon S, Qin D, Manning M, Chen Z, Marquis M, Averyt KB, Tignor M, Miller HL. Cambridge, United Kingdom and New York, NY, USA: Cambridge University Press; 2007:90.

2. van der Werf GR, Morton DC, DeFries RS, Olivier JGJ, Kasibhatla PS, Jackson RB, Collatz GJ, Randerson JT: CO_2 emissions from forest loss. Nature Geoscience 2009, 2: 737–738.

3. UNFCCC: Kyoto Protocol to the United Nations Framework Convention on Climate Change 1998.

4. UNFCCC: The Copenhagen Accord 2009.

5. Cap and Trade California Air Resources Board[http://www.arb.ca.gov/cc/capandtrade/capandtrade.htm]

6. Climate, Community and Biodiversity Project Design Standards Second Edition[http://www.climate-standards.org/]

7. Voluntary Carbon Standard - Guidance for Agriculture, Forestry and Other Land Use Projects[http://www.v-c-s.org/]

8. CAR: Forest Project Protocol, Version 3.2 2010.

9. Winrock International: The American Carbon Registry Forest Carbon Project Standard, Version 2.1 2010.

10. Jenkins JC, Chojnacky DC, Heath LS, Birdsey RA: National-Scale Biomass Estimators for United States Tree Species. Forest Science 2003, 49: 12–35.

11. Smith JE, Heath LS, Jenkins JC: Forest Volume-to-Biomass Models and Estimates of Mass for Live and Standing Dead Trees of U.S. Forests 2003.

12. Jenkins JC, Chojnacky DC, Heath LS, Birdsey RA: Comprehensive Database of Diameter-based Biomass Regressions for North American Tree Species. 2004.

13. Smith JE, Heath LS, Skog KE, Birdsey RA: Methods for Calculating Forest Ecosystem and Harvest Carbon with Standard Estimates for Forest Types of the United States. 2006.

14. McRoberts RE, Nelson MD, Wendt DG: Stratified estimation of forest area using satellite imagery, inventory data, and the k-Nearest

Neighbors technique. Remote Sensing of Environment 2002, 82: 457–468.

15. Hall RJ, Skakun RS, Arsenault EJ, Case BS: Modeling forest stand structure attributes using Landsat ETM+ data: Application to mapping of aboveground biomass and stand volume. Forest Ecology and Management 2006, 225: 378–390.

16. Demaeyer P, De Dapper M, Gamanya : An automated satellite image classification design using object-oriented segmentation algorithms: A move towards standardization. Expert Systems with Applications 2007, 32: 616–624.

17. Song C, Dickinson MB, Su L, Zhang S, Yaussey D: Estimating average tree crown size using spatial information from Ikonos and QuickBird images: Across-sensor and across-site comparisons. Remote Sensing of Environment 2010, 114: 1099–1107.

18. Johansen K, Coops NC, Gergel SE, Stange Y: Application of high spatial resolution satellite imagery for riparian and forest ecosystem classification. Remote Sensing of Environment 2007, 110: 29–44.

19. Koutsias N, Tsakiri-Strati M, Karteris M, Mallinis : Object-based classification using Quickbird imagery for delineating forest vegetation polygons in a Mediterranean test site. ISPRS Journal of Photogrammetry and Remote Sensing 2008, 63: 237–250.

20. Ghioca-Robrecht DM, Johnston CA, Tulbure MG: Assessing The Use Of MultiseasonQuickbird Imagery For Mapping Invasive Species In A Lake Erie Coastal Marsh. Wetlands 2008, 28: 1028–1039.

21. Xiao X, Boles S, Liu J, Zhuang D, Liu M: Characterization of forest types in Northeastern China, using multi-temporal SPOT-4 VEGETATION sensor data. Remote Sensing of Environment 2002, 82: 335–348.

22. Running SW, Nemani RR, Heinsch FA, Zhao M, Reeves MC, Hashimoto H: A continuous satellite-derived measure of global terrestrial primary production. BioScience 2004, 54: 547–560.

23. Grace J, Nichol C, Disney M, Lewis P, Quaife T, Bowyer P: Can we measure terrestrial photosynthesis from space directly, using spectral reflectance and fluorescence? Global Change Biology 2007, 13: 1484–1497.

24. Houghton RA, Butman D, Bunn AG, Krankina ON, Schlesinger P, Stone TA: Mapping Russian forest biomass with data from satellites and forest inventories. Environmental Research Letters 2007, 7.

25. Potter C, Klooster S, Huete A, Genovese V: Terrestrial carbon sinks for the United States predicted from MODIS satellite data and ecosystem modeling. Earth Interactions 2007, 11: 1–21.

26. Zheng D, Heath LS, Ducey MJ: Forest biomass estimated from MODIS and FIA data in the Lake States: MN, WI and MI, USA. Forestry 2007, 80: 265–278.

27. Baccini A, Laporte N, Goetz SJ, Sun M, Dong H: A first map of tropical Africa's above-ground biomass derived from satellite imagery. Environmental Research Letters 2008, 3: 9.

28. Blackard JA, Finco MV, Helmer EH, Holden GR, Hoppus ML, Jacobs DM, Lister AJ, Moisen GG, Nelson MD, Riemann R, Ruefenacht B, Salajanu D, Weyermann DL, Winterberger KC, Brandeis TJ, Czaplewski RL, McRoberts RE, Patterson PL, Tymcio RP: Mapping U.S. forest biomass using nationwide forest inventory data and moderate resolution information. Remote Sensing of Environment 2008, 112: 1658–1677.

29. Goetz S, Baccini A, Laporte N, Johns T, Walker W, Kellndorfer J, Houghton R, Sun M: Mapping and monitoring carbon stocks with satellite observations: a comparison of methods. Carbon Balance and Management 2009, 4: 2.

30. Paivinen R, Van Brusselen J, Schuck A: The growing stock of European forests using remote sensing and forest inventory data. Forestry 2009, 82: 479–490.

31. Lefsky MA, Cohen WB, Parker GG, Harding DJ: Lidar Remote Sensing for Ecosystem Studies. BioScience 2002, 52: 19–30.

32. Lefsky MA, Hudak AT, Cohen WB, Acker SA: Patterns of covariance between forest stand and canopy structure in the Pacific Northwest. Remote Sensing of Environment 2005, 95: 517–531.

33. Balzter H, Rowland CS, Saich P: Forest canopy height and carbon estimation at Monks Wood National Nature Reserve, UK, using dual-wavelength SAR interferometry. Remote Sensing of Environment 2007, 108: 224.

34. Wallerman J, Holmgren J: Estimating field-plot data of forest stands using airborne laser scanning and SPOT HRG data. Remote Sensing of Environment 2007, 110: 501–508.

35. Hilker T, Wulder MA, Coops NC: Update of forest inventory data with lidar and high spatial resolution satellite imagery. Canadian Journal of Remote Sensing 2008, 34: 5–12.

36. Ioki K, Imanishi J, Sasaki T, Morimoto Y, Kitada K: Estimating stand volume in broad-leaved forest using discrete-return LiDAR: plot-based approach. Landscape EcolEng 2009, 6: 29–36.

37. Gonzalez P, Asner GP, Battles JJ, Lefsky MA, Waring KM, Palace M: Forest carbon densities and uncertainties from Lidar, QuickBird, and field measurements in California. Remote Sensing of Environment 2010, 114: 1561–1575.

38. Takahashi T, Awaya Y, Hirata Y, Furuya N, Sakai T, Sakai A: Stand volume estimation by combining low laser-sampling density LiDAR - data with QuickBird panchromatic imagery in closed-canopy Japanese cedar (Cryptomeria japonica) plantations. International Journal of Remote Sensing 2010, 31: 1281.

39. Straub C, Weinacker H, Koch B: A comparison of different methods for forest resource estimation using information from airborne laser scanning and CIR orthophotos. Eur J Forest Res 2010, 129: 1069–1080.

40. Breidenbach J, Næsset E, Lien V, Gobakken T, Solberg S: Prediction of species specific forest inventory attributes using a nonparametric semi-individual tree crown approach based on fused airborne laser scanning and multispectral data. Remote Sensing of Environment 2010, 114: 911–924.

41. Magnussen S, Boudewyn P: Derivations of stand heights from airborne laser scanner data with canopy-based quantile estimators. Can J For Res 1998, 28: 1016–1031.

42. Akay A, Oğuz H, Karas I, Aruga K: Using LiDAR technology in forestry activities. Environmental Monitoring and Assessment 2009, 151: 117–124.

43. Ståhl G, Holm S, Gregoire TG, Gobakken T, Næsset E, Nelson R: Model-based inference for biomass estimation in a LiDAR sample survey in Hedmark County, Norway. Can J For Res 2011, 41: 96–107.

44. Maltamo M, Bollandsås OM, Næsset E, Gobakken T, Packalén P: Different plot selection strategies for field training data in ALS-assisted forest inventory. Forestry 2011, 84: 23–31.

45. Coops NC, Wulder MA, Culvenor DS, St-Onge B: Comparison of forest attributes extracted from fine spatial resolution multispectral and lidar data. Canadian Journal of Remote Sensing 2004, 30: 855–866.

46. Wang Y, Weinacker H, Koch B: A Lidar Point Cloud Based Procedure for Vertical Canopy Structure Analysis And 3D Single Tree Modelling in Forest. Sensors 2008, 8: 3938–3951.

47. Maltamo M, Peuhkurinen J, Malinen J, Vauhkonen J, Packalén P, Tokola T: Predicting tree attributes and quality characteristics of Scotspine using airborne laser scanning data. Silva Fennica 2009, 43: 507–521.

48. Næsset E: Predicting forest stand characteristics with airborne scanning laser using a practical two-stage procedure and field data. Remote Sensing of Environment 2002, 80: 88–99.

49. Næsset E: Estimating timber volume of forest stands using airborne laser scanner data. Remote Sensing of Environment 1997, 61: 246–253. 1

50. Næsset E: Determination of mean tree height of forest stands using airborne laser scanner data. ISPRS Journal of Photogrammetry and Remote Sensing 1997, 52: 49–56.

51. Hudak AT, Crookston NL, Evans JS, Falkowski MJ, Smith AMS, Gessler PE, Morgan P: Regression modeling and mapping of coniferous forest basal area and tree density from discrete-return lidar and multispectral satellite data. Canadian Journal of Remote Sensing 2006, 32: 126–138.

52. Hudak A, Evans JS, Crookstone NL, Falkowski MJ, Steigers BK, Taylor R, Hemingway H: Aggregating Pixel-Level Basal Area Predictions Derived from LiDAR Data to Industrial Forest Stands in North-Central Idaho. USDA Forest Service 2008, 14.

53. Hudak AT, Crookston NL, Evans JS, Hall DE, Falkowski MJ: Nearest neighbor imputation of species-level, plot-scale forest structure attributes from LiDAR data. Remote Sensing of Environment 2008, 112: 2232–2245.

54. Nothdurft A, Saborowski J, Breidenbach J: Spatial prediction of forest stand variables. Eur J Forest Res 2009, 128: 241–251.

55. Breidenbach J, Nothdurft A, Kändler G: Comparison of nearest neighbour approaches for small area estimation of tree species-specific forest inventory attributes in central Europe using airborne laser scanner data. Eur J Forest Res 2010, 129: 833–846.

56. Latifi H, Nothdurft A, Koch B: Non-parametric prediction and mapping of standing timber volume and biomass in a temperate forest: application of multiple optical/LiDAR-derived predictors. Forestry 2010, 83: 395–407.

57. Jaskierniak D, Lane PNJ, Robinson A, Lucieer A: Extracting LiDAR indices to characterise multilayered forest structure using mixture distribution functions. Remote Sensing of Environment 2011, 115: 573–585.

58. Packalén P, Maltamo M: Predicting the Plot Volume by Tree Species Using Airborne Laser Scanning and Aerial Photographs. Forest Science 2006, 52: 611–622.

59. Moeur M, Stage AR: Most Similar Neighbor: An Improved Sampling Inference Procedure for Natural Resource Planning. Forest Science 1995,41(23):337–359.

60. Lindberg E, Holmgren J, Olofsson K, Wallerman J, Olsson H: Estimation of tree lists from airborne laser scanning by combining single-tree and area-based methods. International Journal of Remote Sensing 2010, 31: 1175.

61. Aalde H, Gonzalez P, Gytarsky M, Krug T, Kurz WA, Ogle S, Raison J, Schoene D, Ravindranath NH, Elhassan NG, Heath LS, Higuchi N, Kainja S, Matsumoto M, Sanchez MJS, Somogyi Z: Chapter 4: Forest Land. In 2006 IPCC Guidelines for National Greenhouse Gas Inventories. Edited by: HS E, LB, K M, T N, KT Iges. Japan: National Greenhouse Gas Inventories Programme; 2006.

62. Hawbaker TJ, Keuler NS, Lesak AA, Gobakken T, Contrucci K, Radeloff VC: Improved estimates of forest vegetation structure and biomass with a LiDAR-optimized sampling design. J Geophys Res 2009, 114: 11.

63. Gobakken T, Naesset E: Assessing effects of laser point density, ground sampling intensity, and field sample plot size on biophysical stand properties derived from airborne laser scanner data. Canadian Journal of Forest Research 2008, 38: 1095–1109.

64. van Aardt JAN, Wynne RH, Oderwald RG: Forest Volume and Biomass Estimation Using Small-Footprint Lidar-Distributional Parameters on a Per-Segment Basis. Forest Science 2006, 52: 636–649.

65. Avery TE, Burkhart HE: Forest Measurements. 5th edition. New York, NY: McGraw-Hill Higher Education; 2002.

66. Shiver BD, Borders BE: Sampling techniques for forest resource inventory. New York, NY: John Wiley & Sons, Inc; 1996.

67. Borders BE, Shiver BD, Clutter ML: Timber Inventory of Large Acreages Using Stratified Two-Stage List Sampling. Southern Journal of Applied Forestry 2005, 29: 152–157.

68. Bell JF, Dilworth JR: Log Scaling and Timber Cruising. 2007 edition. Corvallis, OR: Cascade Printing Company; 2007.

69. Clark ML, Roberts DA, Clark DB: Hyperspectral discrimination of tropical rain forest tree species at leaf to crown scales. Remote Sensing of Environment 2005, 96: 375–398.

70. Levy PS, Lemeshow S: Sampling of Populations: Methods and Applications. 4th edition. Wiley; 2008.

71. GPSMAP 76CSx Owner's Manual. 1200 East 151st St. Olathe, Kansas 66062 Garmin;

72. Pesonen A, Kangas A, Maltamo M, Packalén P: Effects of auxiliary data source and inventory unit size on the efficiency of sample-based coarse woody debris inventory. Forest Ecology and Management 2010, 259: 1890–1899. 10.1016/j.foreco.2010.02.0

73. Lumley T, Miller A: leaps: regression subset selection. 2009.

74. R Development Core Team: R: A Language and Environment for Statistical Computing. Vienna, Austria: R Foundation for Statistical Computing; 2011.

75. Ramsey FL, Schafer DW: The Statistical Sleuth: A Course in Methods of Data Analysis. 2nd edition. Pacific Grove, CA, USA: Duxbury Thomson Learning; 2002.

76. California Department of Forestry and Fire Protectiton: Important Information for Timber Operations Proposed with the Range of Northern Spotted Owl Sacramento, CA: Department of Forestry and Fire Protection; 2008:35.

77. The Conservation Fund: Garcia River Forest Integrated Resource Management Plan Caspar, CA: The Conservation Fund; 2006:289.

78. MATLAB Image Processing Toolbox Natick, Massachusetts, U.S.A: Mathworks; 2011.

79. Gonzalez RC, Woods RE: Digital Image Processing. 2nd edition. Upper Saddle River, NJ: Prentice Hall; 2002.

80. Jolliffe IT: Principle Component Analysis. 2nd edition. Springer-Verlag; 2002.

81. Harrell FE: Regression Modeling Strategies. Corrected. Springer; 2001

82. Derksen S, Keselman HJ: Backward, forward and stepwise automated subset selection algorithms : frequency of obtaining authentic and noise variables. British Journal of Mathematical and Statistical Psychology 1992, 45: 265–282.

83. Altman DG, Andersen PK: Bootstrap investigation of the stability of a Cox regression model. Stat Med 1989, 8: 771–783.

84. Tibshirani R: Regression Shrinkage and Selection via the Lasso. Journal of the Royal Statistical Society. Series B (Methodological) 1996, 58: 267–288.

85. Glover F: Tabu Search: A Tutorial. Interfaces 1990, 20: 74–94.

86. Glover F, Taillard E, de Werra D: A user's guide to tabu search. Ann Oper Res 1993, 41: 1–28.

87. Thompson SK: Sampling. 2nd edition. New York, NY: John Wiley & Sons, Inc; 2002.

88. United States Forest Service Pacific Northwest Research Station: Volume estimation for the PNW-FIA Database. USFS Pacific Northwest Research Station: United States Department of Agriculture, Forest Service. 2010, 69.

CHAPTER 3

Detection and Classification of Changes in Buildings from Airborne Laser Scanning Data

Sudan Xu *, George Vosselman and Sander Oude Elberink

Department of Earth Observation Science, Faculty ITC, University of Twente, PO BOX 217, 7500 AE Enschede, The Netherlands

ABSTRACT

The difficulty associated with the Lidar data change detection method is lack of data, which is mainly caused by occlusion or pulse absorption by the surface material, e.g., water. To address this challenge, we present a new strategy for detecting buildings that are "changed", "unchanged", or "unknown", and quantifying the changes. The designation "unknown" is applied to locations where, due to lack of data in at least one of the epochs, it is not possible to reliably detect changes in the structure. The process starts with classified data sets in which buildings are extracted. Next, a point-to-plane surface difference map is generated by merging and comparing the two data sets. Context rules are applied to the difference map to distinguish between "changed", "unchanged", and "unknown". Rules are defined to solve problems caused by the lack of data. Further, points labelled as "changed" are re-classified into changes to roofs, walls, dormers, cars, constructions above the roof line, and undefined objects. Next, all the classified changes are organized as changed building objects, and the geometric indices are calculated from their 3D minimum bounding boxes.

Performance analysis showed that 80%–90% of real changes are found, of which approximately 50% are considered relevant.

KEYWORDS

building; change; classification; ALS; change detection; airborne laser scanning

1. INTRODUCTION

In urban areas, changes to buildings may be caused by natural disasters or geological deformation, but more often they are the result of human activities. These activities may lead to temporary or permanent changes, as, for example, discussed by Xiao et al. [1]. Detection of structural changes to urban objects, e.g., renovation of infrastructural objects and buildings, is important for municipalities, which need to keep their topographic object databases up-to-date.

Analysis of the geometric differences between point clouds from different epochs of data provides an insight into how areas change over time. The general approach in using point clouds for change detection is to focus on areas where points are present in one epoch of data and absent—at least in the close vicinity—in the other. As the need for detecting changes at higher levels of detail is growing, demand for more detailed interpretation of differences between epochs of data has also grown. There can be several reasons for absence of data points for a certain location in one epoch of data even if data points for that same location are present in another epoch, for example:

1) the area around the location may have been occluded in one of the data epochs;
2) the surface around the location may have absorbed the ALS laser pulses;
3) outliers (either on the ground or inside buildings) may be present; or
4) the object may have undergone change.

In the first three cases, differences in the data do not correspond with real changes. In the fourth (last) case, some of the changes may not be relevant for a municipality's databases, e.g., a parking place with a lot of changes due to the temporary presence or absence of cars, which results in different arrangements of point clouds. The main problem is that it is not known whether a difference between two data sets is caused by differences in scanning geometry, surface properties, or changes to an object—relevant or irrelevant. A better understanding of differences between epochs of data would allow us to precisely determine whether there has been a relevant change in an area and, if so, what kind of change it is.

Our study aims to develop a method for detecting relevant changes occurring in urban objects using point clouds from ALS data. Our focus is on changes to buildings, including changes to roof elements and those associated with car parking lots on top of buildings. The main challenge is how to separate irrelevant differences from relevant object changes. We consider changes on objects with an area of <4 m^2 to be irrelevant changes (although we can detect them), because we assume that 4 m^2 is the minimum area that can accommodate a person for the purpose of checking the building or property.

With the above aims in mind, we set up the following change detection procedure: first, in each point cloud, objects are classified as "ground", "water", "vegetation", "building roof", "roof element", "building wall", or "undefined object" using a rule based classifier with different entities as seen in Xu et al. [2]. This classification step is referred to as "scene classification" in the remainder of this paper. Next, a surface difference map is generated by calculating the point-to-plane distances between the points in one epoch of data to their nearest planes in the other epoch. By combining scene classification information with the surface difference map, changes to building objects can be detected. Points on these changed objects are grouped and further analyzed in a rule-based context to classify them as changes related to: a roof, a wall, a dormer, cars on flat roofs, a construction on top of a roof, or undefined objects. Reference data was collected manually to determine the accuracy of this approach.

To begin, we describe the state-of-the-art in change detection techniques in Section 2. The explanation of the method in Section 3 looks at our use of surface difference maps (Section 3.1), our change detection algorithm (Section 3.2), the classification of changes (Section 3.3), and our analysis of object-based changes (Section 3.4). The data sets used are described in Section 4, followed by presentation of results and their analysis in Section 5. We offer our conclusions in Section 6.

2. RELATED RESEARCH

Research related to change detection is discussed in two parts. First, an overview is given of previous research done on change detection from imagery. In the second part, we discuss techniques using lidar point clouds for detecting changes in buildings.

2.1. Change Detection from Imagery

According to Mas [3], change detection from imagery can be grouped into three categories divided by the data transformation procedures and the analysis techniques: (1) image enhancement; (2) multi-date data classification; and (3) comparison of two independent land cover classifications. The first group of image enhancement includes image differencing, image regression, image rationing, vegetation index differencing, change vector analysis (CVA), IRMAD, new kernel based methods, and transformation methods such as Principle Component Analysis (PCA) [4] and RANSAC [5]. The main problem of all the algorithms is to determine the threshold for the change areas. Among all these methods, CVA performs best. CVA is an extension of image differencing and it computes the change direction and magnitude of all bands for all dates that the images were taken.

The multi-date data classification approach is based on pixels being directly classified as changed or unchanged according to differences distinguished between features of the pixels and/or those of their neighborhoods [6]. The third approach is based on comparing two independent classified images and is also called a post-classification comparison [3]. Both images are classified independently and differences in pixel labels are assumed to be caused by changes in, for example, land cover [7]. Obviously, the quality of the change detection depends on the quality of the classification. Classification methods can be supervised and unsupervised. Yang and Zhang [6] and Di et al. [7] both used the Support Vector Machine (SVM) as a supervised classifier. Unsupervised methods have been described by Tanathong et al. [8], who defined a classifier agent and an object agent, and Kasetkasem and Varshney [9], who introduced Markov Random Field Models for change detection.

2.2. Change Detection in Lidar Point Clouds

Similar to image differencing, the first approach used for detecting change in multi-temporal lidar point clouds involved the subtraction of two DSMs [10] from each other. Classification is also used to directly distinguish changed objects from unchanged ones; this is usually applied in disaster assessment. For an example of this method of classification, see Khoshelham et al. [11], who, using an ALS data set, performed several supervised classifiers on a small number of training samples to distinguish damaged roofs from undamaged roofs.

In cases of 3D change detection in buildings, depending on the nature of the application and the availability of data sets, comparisons can be made between data sets of multi-temporal images, between multi-epoch lidar point clouds, and between image and lidar point clouds. With the improvements in the accuracy of obtained images as well as improvements in processing skills, 3D point clouds can also be derived from images by, for example, dense matching algorithms [12].

There are two basic approaches to the problem of change detection, each determined by the availability of original data: (1) original data is available from both epochs to be used for detecting change in cases of disasters or geological deformation; (2) original data is only available for one epoch, while the other epoch of data is an existing map or database [13]. The literature of these two approaches is reviewed below:

Approach 1

In 1999, Murakami et al. [10] used two ALS data sets, one acquired in 1998 and the other in 1996, to detect changes to buildings after an earthquake in a dense urban area in Japan. A difference map was obtained by subtracting one DSM from the other. This difference map was laid over an ortho-image and an existing GIS database to identify changes to buildings. Vögtle and Steinle [14] presented their research as a part of a project using DSMs from two ALS data sets to detect changes after strong earthquakes. Segmentation was run to find all the buildings in the area. The changes were identified by the overlay rate of all the buildings in both epochs of data. Rutzinger et al. [15] extracted buildings from DSMs of two data epochs using a classification tree. Shape indices and the mean height difference of the building segments were compared. Differences between classified DSMs derived from two epochs of ALS data have also been used by Choi et al. [16] to detect changes.

Approach 2

Vosselman et al. [13] compared ALS data to an existing medium-scale map. Change detection was done by segmentation, classification, and the implementation of mapping rules. Pixel overlay rates on a classified DSM and a raster map were used to finally identify changes. This method was then improved by using aerial images to refine the classification results [17]. Rottensteiner [18] employed data fusion with the Dempster-Shafer theory for building detection. He then improved his method by adding one more feature to the data fusion, which makes the classification suitable for building-change detection. Champion et al. [19] detected building changes by comparing a DSM with a vector map. The similarity measure between building outlines in the vector map and the DSM contours was used to identify the demolished, modified, and unchanged buildings. New buildings were detected separately in the DSM. Chen et al. [20] used lidar data and an aerial image to update old building models. After registration of the data, the area of change was detected and height differences were calculated between the roof planes in the lidar data and planes in the old building models. A double-thresholding strategy was used to identify the main-structures in changed and unchanged areas, while uncertain parts were identified from the line comparisons between building boundaries extracted from an aerial image and the projection of the old building models. The double-thresholding method was reported to improve overall accuracy from 93.1% to 95.9%. Awrangjeb et al. [21] detect building changes from lidar point cloud by determining the extended or demolished part of a building with a connected component analysis and they conclude that no omission errors are found.

2.3. Problems of Existing Methods and Our Contribution

Approaches described above enable changes to be detected in 2D (maps) or 2.5D (DSMs), which may result in information loss such as changes under tall trees. To avoid these problems, methods that can compare the lidar data directly are required. In 2011, Hebel et al. [22] introduced an occupancy grid to track changes explicitly in multi-temporal 3D ALS data. In order to analyze all the laser beams passing through the same grid, they defined belief masses "empty", "occupied", and "unknown" for each voxel in the object space. For the data sets they compared, they computed belief masses resulting from all the laser beams, with conflicts in belief masses denoting a change. By adding an extra attribute to indicate the smoothness and continuity of a surface, they were able to achieve reliable change detection results, even when occlusion had occurred in either of the data epochs.

Subsequently, Xiao et al. [1] introduced an occupancy grid to the mobile laser scanning system to detect permanent objects in street scenes, where occlusions frequently occur. They too reported that occupancy grids are resistant to occlusion. However, when point density is too low, for example on walls, and the size of the occupancy grid is small, parts of walls were identified as occluded because no occupied points were included in the grid due to the low point density. The size of the occupancy grid may, therefore, influence the detection result when the point density of the lidar data varies from place to place. Besides the occlusion problem, pulse absorption may cause geometric differences in the data, although no real change occurred, as explained in the introduction (Section 1).

Compared to the existing methods, our contribution consists of: (1) avoiding information loss by directly comparing the lidar data (not DSM) using the surface difference map when changes occur under vegetation; (2) solving problems of occlusion and pulse absorption on building tops without using occupancy grids; (3) evaluating the accuracy per changed object, be it just a small dormer or a large building (or part of a building).

3. METHODOLOGY

Our approach can be labelled as a post-classification comparison on multi-epoch 3D point clouds. To interpret changes in a scene, we start by calculating the differences between two data sets. A surface difference map is generated by calculating the point-to-plane distance between the points from one data epoch to their nearest planes in the other (see Section 3.1 for more details). As the points on building roofs, walls, and roof elements are considered to be extracted in the scene classification step, it is possible to combine the scene classification result with surface difference information. This combination enables us to not only detect changes on building objects but also detect occluded areas, where it is not known whether there has been a change or not (see Section 3.2). Points on the changed objects are grouped and analyzed according to a rule-based context. In addition to changes to roofs and walls (see Section 3.3), changes to roof elements are further classified as cars, construction, dormers, or undefined objects. The classified changes are finally grouped into building objects and analyzed (see Section 3.4). Reference data were collected manually at random locations to determine the accuracy of our method.

3.1. Generating a 3D Surface Difference Map

The 3D surface difference map records the disparities of points between two epochs of ALS data. The disparity per point is computed as the distance from a point to its nearest fitted plane from another epoch. Surface difference was employed by Vosselman [23], who evaluated the quality of data using overlapping strips. In our paper, the surface difference map is used to indicate 3D differences between two lidar data sets.

For every point in one epoch of data, we search within its 1.0 m (radius) 3D neighborhood to check whether there is a point from the other epoch. If there are no points from the other epoch, this implies a difference greater than 1.0 m. In these cases we record a difference value >1.0 m. If there are points from the other epoch of data, we define the surface difference as the distance from the selected point to the nearest fitted plane in the point cloud of the other epoch.

A 3D neighborhood of 1.0 m radius points is chosen because the point density is greater than 1 point/m^2 in most locations of our dataset. This ensures that the comparison is not affected by the point density. Although we have two data sets that have similar point densities, in order to make sure that a 1.0 m neighborhood is suitable for comparisons between data sets with different point densities. We simulated several data sets with our test data by reducing the point density of one of the epochs (see Figure 1).

The difference map contains the geometric indication of whether there is a change or not. However, not every difference is a change (Figure 2), and sometimes part of a change is not represented as a large value in the difference map (Figure 3). Figure 2c shows the difference map containing several points with a surface difference greater than 10 cm. However, these differences are caused by lack of data in the other epoch and therefore cannot be considered as a change in an object. Conversely, points with a difference value less than 10 cm may belong to a changed object. Figure 3 is an example that displays points with a surface difference of less than 10 cm in the connections between a dormer and a roof; the change was detected because the dormer was newly built.

Our task is to assign the labels of “changed”, “unchanged”, and “unknown” to each point according to the surface difference map.

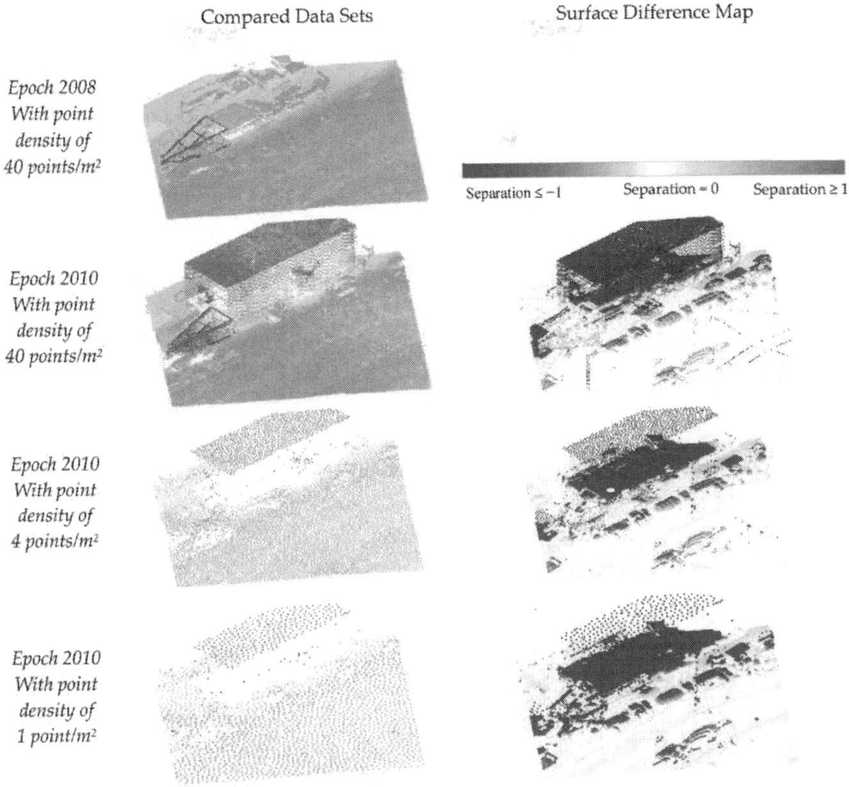

Figure 1. Examples of the surface difference map for various point densities. Comparison of data sets with different densities from the year 2010 with the data from the year 2008 showed that the locations of the changes are the same in all the difference maps. This indicates that the surface difference map generated with a neighborhood of 1.0 m points is suitable for all the data sets that have a point density larger than 1 point/m^2.

3.2. Detecting a Change

The strategy for interpretation of the surface difference map is as follows: points classified as building (wall, roof, or roof element) in the scene classification are selected. All building points are assumed to be unchanged except for those distinguished as "unknown" or "changed".

Points with a difference value greater than 1 m in one epoch for which, even in a two dimensional neighborhood, no nearby points can be found in the other epoch are considered "unknown" (see Figure 4) . This occurs in areas where there is a lack of data in one epoch because of occlusions or pulse absorption by the surface. For walls, large surface difference values may occur. However, our algorithm will label the points for the wall as "unknown" instead of "changed" if there has been no change to the roof of that same building. Due to a lack of evidence as to what happened with the wall, the points are labelled as "unknown".

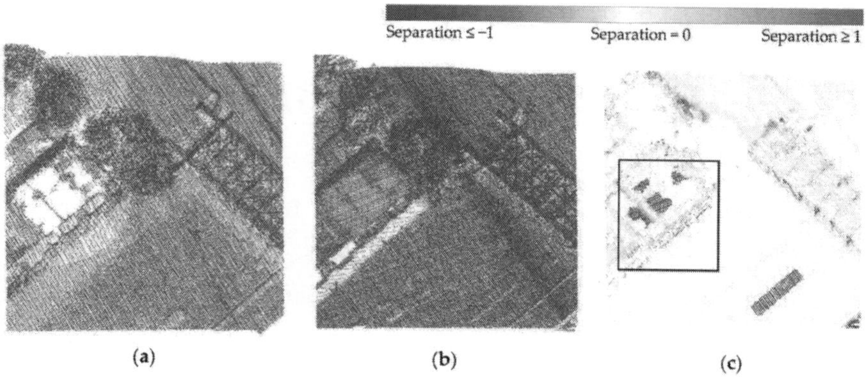

Figure 2. (**a**,**b**): Different colors represent different epochs of data, (**a**) 2008, (**b**) 2010; (**c**) Surface difference map derived from the two epochs. Points with a great difference do not always denote a change. For example, in (**c**) blue points in the rectangle are unknown points because there is no data for the roof in thr 2008 epoch data (perhaps caused by water on the roof), although they have high difference values.

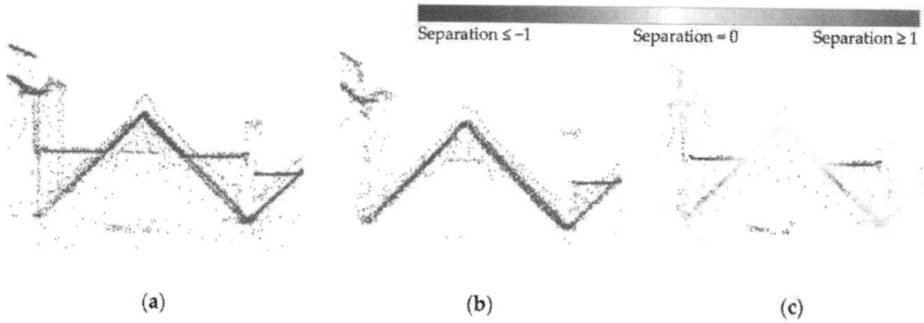

Figure 3. (**a**,**b**) Different colors represent different epochs of data, (**a**) 2012, (**b**) 2010; (**c**) Surface difference map—both epochs merged. The points in the rectangle have a distance difference ≤0.08 m, although there is indeed a change. The difference value is small because the points on the roof from the other epoch are in the neighborhood of the points on the dormer and they are quite close to each other. The problem is to identify the areas in the rectangle as being changed when the difference value is low.

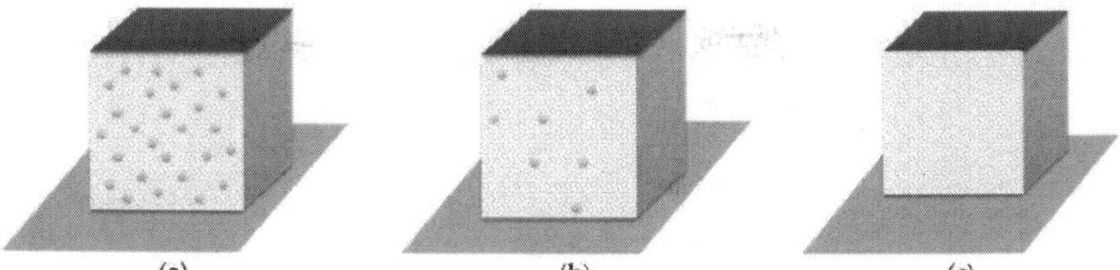

Figure 4. Points labelled as “unknown” for walls. (**a**) a wall scanned with dense points; (**b**) the same wall scanned with sparse or (**c**) no points in another epoch.

In Table 1, the “seed neighborhood radius” and the “growing radius” indicate the maximum radius for allowing a point participating in the seed plane extraction and the plane growing phase. The “minimum number of seed points” means the minimum number of points that are required for initialization of a plane. The plane parameters are obtained from the Hough space and 10 points are required as the minimum number of seed points to ensure a reliable plane extraction. The “maximum distance of a point to the plane” ensures that only nearly co-planar points are added to the seed plane. The 1.0 m radius is set to ensure that the surface can be extended with neighbouring points on the same surface. This setting needs to be adjusted for processing datasets with a different point density.

Generally, “changed” points have a high difference value, although they cannot be selected using a simple threshold in the difference map. Therefore, we derived a rule-based decision-tree to detect “changed” points in surface difference maps. The decision-tree is shown in Figure 5. The “unknown” points are first excluded from the data sets. The remaining points are grouped into planar segments using the surface growing method [24]. Within each segment, points are further separated into connected components with a smaller radius than during the planar surface growing step. This separates two nearby objects that are located in same plane. For example, Figure 6a shows two dormers belonging to the same planar segment, and after deriving connected components, they are separated into two components. These components are used as the basic units for identifying changed points. Parameters used in the surface growing and connected component algorithms are listed in Table 1.

If, for each connected component, the vast majority of points have surface differences greater than the maximum strip difference, the whole component is very likely to be a change and will be labelled as “changed”. Otherwise, the component may be unchanged or only partly changed. If partly changed, the changed part is only labeled as “changed” by checking its area. The parameters for the vast majority, the maximum strip difference, and the area are listed in Table 2 and the reason for the parameter setting is given.

Table 1. Parameters for the surface growing and connected component algorithm.

Surface Growing				Connected Component
Parameters for seed selection		Parameters for growing		Maximum distance of component
Seed neighborhood radius	1.0 m	Growing radius	1.0 m	1.0 m
Minimum number of seed points	10 pts	Maximum distance of a point to the plane	0.1 m	

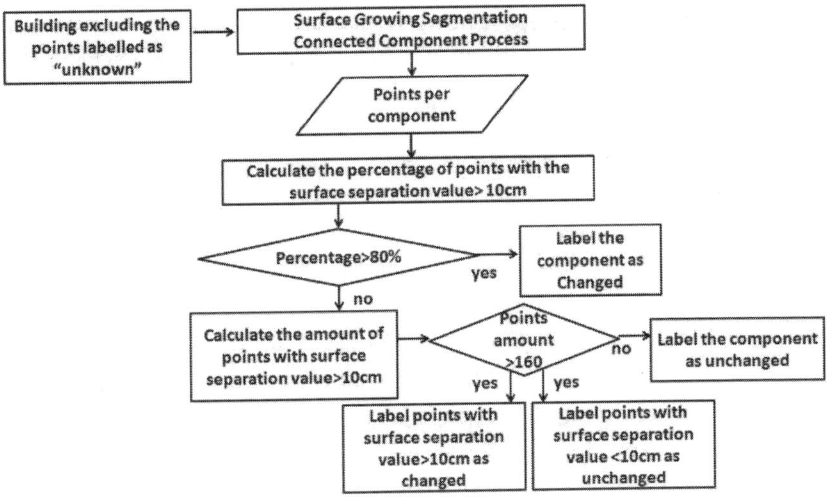

Figure 5. Identification of changed and unchanged points using a decision-tree.

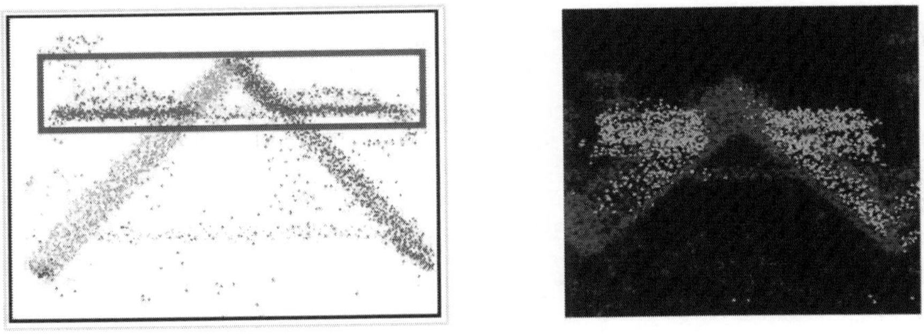

(a) (b)

Figure 6. (**a**) The two dormers in the red rectangle belong to the same planar segment and therefore need connected components analysis to separate them (different colors indicate different segments); (**b**) Both changes to roofs and dormers are detected (two data epochs are merged).

3.3. Classification of Building Change

Points detected as changes in a building are extracted and a second classification step is performed on these changes. This second classification step is required to understand the activities that could have possibly led to the changes detected for a building. For example, by identifying changes to roof elements on top of a building (Figure 7a), we can infer that these elements may be cars, and that there is a parking lot on top of the building.

Table 2. Parameters for deciding on change.

Name	Thresholds	Reason for the Choice of Thresholds	Suitability to Transfer to Other Datasets
Vast majority	80%	Tested	To be adjusted according to the training data.
Maximum strip height difference between two epochs of data	10 cm	Registration error in each epoch was shown to be below 5 cm.	To be adjusted to the maximum registration error.
Area	160 points (4 m^2)	The average point density is 40 pts/m^2, and all buildings larger than 4 m^2 should be mapped.	To be adjusted to the point density and the definition of relevant change.

Figure 7. Connected components of changed objects are classified based on their context. (**a**) 2008 vs. 2010; (**b**) 2010 vs. 2012.

There are various types of objects that may be present on building roofs or attached to walls. Most changes to walls are caused by sun shades (which are open in one epoch but closed in the other epoch), stairs, flags, or vegetation near walls. Compared to the changes to roofs (e.g., a new dormer or the addition of a floor), changes near walls are less likely to be related to construction activities. For this reason, the second classification step is performed only for changes concerning roofs and roof elements.

Based on the size of changes and the underlying building structure, changes on a roof top are classified as dormers, cars or “other constructions”. The difference between a dormer and “other constructions” above the roof line is that the latter involves relatively large changes compared to a dormer. The attributes used for the classification are area, height to the nearest roof, the normal vector direction of the nearest roof (which indicates a pitched or a flat roof) and the labels from the scene classification results.

Rules are defined to classify these components. Large areas of points labelled as changed indicate a change to a roof. Small changes occurring on a pitched roof are more likely to be newly built or removed dormers or chimneys. If changes occur on a flat roof, they are probably constructions above the roof line or cars. These rules are used in a decision-tree, and have been established as a rule-based classifier with some defined values according to the knowledge (see Figure 8). All defined values are based on the authors’ knowledge and are explained in Table 3.

Table 3. Defined parameter values for the rule based classifier.

Attribute Name	Value	Explanation
Approximate area		
On pitched roof On flat roof	$4\ m^2$ $6\ m^2$	Minimum area of interest Area of small vehicle
Height to the nearest roof	3 m	The maximum height of a car
Class label of point in the scene classification	"roof"/"roof element"/"wall"	For understanding the context

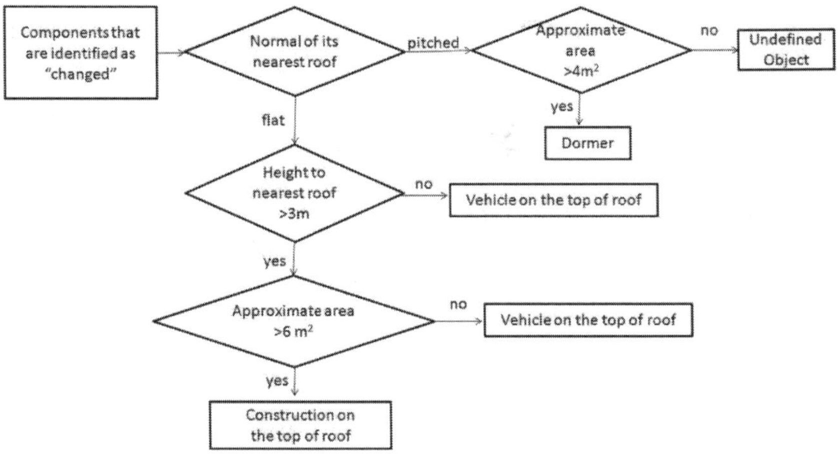

Figure 8. Rule-based classification of changes.

3.4. Analysis of Changed Building Objects

After the points are identified as "changed" and classified, the results are points organized as connected components. These components have not been identified as building objects yet, and they may represent only a small part of an object. How points are organized as building objects and how the changed objects are analyzed is described in the following subsections.

3.4.1. Changed Building Objects

The points that are identified as changed can be organized into building objects once we have labelled all the points with: (1) the type of object class from the change classification; (2) the labels "changed", "unchanged", or "unknown"; and (3) the signs from a difference map indicating a newly built or demolished structure. We distinguish buildings as being entirely changed or partly changed.

If there is a change to the roof of a building, all points that group together in a 2D connected component step will be labelled as an entirely changed building. If changes are made to building elements, the local 3D connected points will be only labelled as partly changed.

3.4.2. Merging Objects at Tile Boundaries

So far, changes are detected within tiles of, for example, 50 m × 50 m. Typically, buildings are stored in different tiles, as shown in Figure 9a–c. As changed objects may occur on tile boundaries, a merging step is needed to detect a single object instead of two (i.e., one in each neighbouring tile). To solve this problem, adjacent components are merged across tile boundaries, as described by Vosselman [25]. Figure 9d shows the components after they have been merged.

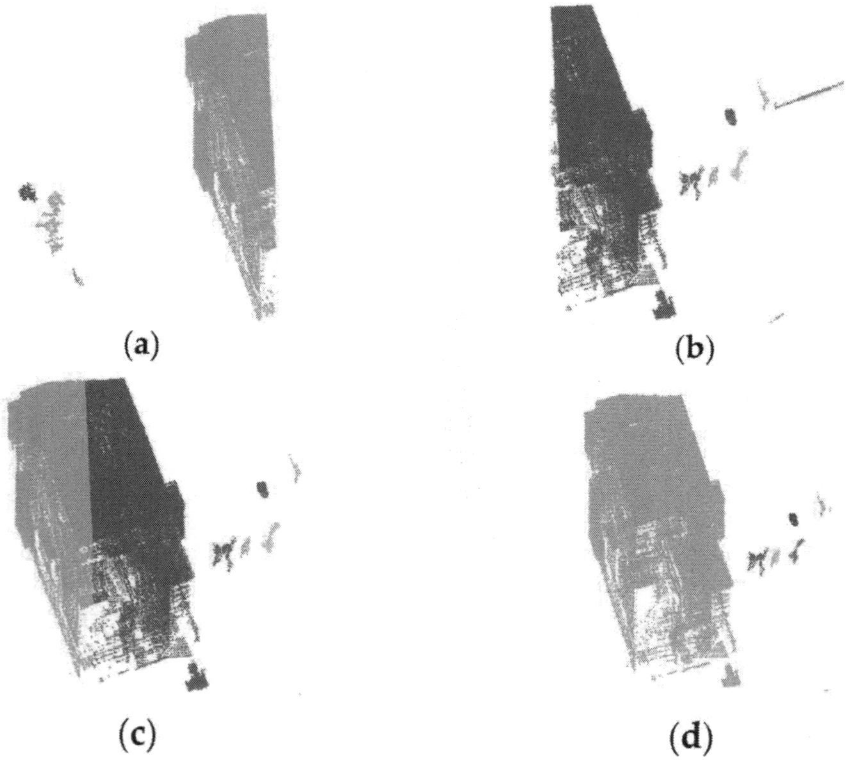

Figure 9. Objects are cut and stored in different tiles (**a**) and (**b**) during the processing of large data sets. Components have been merged in (**c**) and (**d**). (**a**) building part in one tile; (**b**) building part in another tile; (**c**) the same building part belonging to different objects; (**d**) the same building part merged into the same object.

When building objects have been formed, minimum 3D bounding boxes are calculated for these objects. Their width, length, height, area, volume, and location of their center point are determined from their 3D bounding boxes.

Among all the changes near buildings we found, there are some that are irrelevant for the municipality. Common examples are changes to extensive areas of flowers and shrubs, to fences, and railings in gardens on rooftops of buildings. Bounding boxes have been used to measure more precisely the size of a change. This measure of size is used to separate irrelevant changes (<4 m^2) from relevant ones. For changes <4 m^2 and a "length: width" ratio greater than 10, no 3D bounding boxes are calculated.

4. DATA SETS

The research we present here originated from a request from the Municipality of Rotterdam (the Netherlands) to help in monitoring changes to buildings using ALS data. Three data sets, located in commercial and residential areas of Rotterdam, have been used. The data of the commercial area was scanned in years 2008 and 2010; data sets of the residential area were acquired in years 2010 and 2012 (Figure 10). The point densities for the data sets 2008, 2010, and 2012 are average 30 points/m^2, 35 points/m^2 and 40 points/m^2, respectively. The strip difference within an epoch and between different epochs had a maximum value of 10 cm. As the data sets already register well (with a strip difference less than 10 cm), no extra steps for registration were required. All data sets underwent scene classification and were then divided into tiles of 50 m × 50 m. Figure 11 shows four examples of the result from the scene classification step. Bridges and its cables are classified as building roof and roof elements in data set 2008 and 2010. Data set 2010 was obtained in the fall and the 2012 dataset was obtained in the fall. Therefore, tree points are more dense in the 2012 data set.

5. RESULT AND DISCUSSION

Our change detection procedure comprises change detection, classification of any change detected, and object-based analysis. This procedure includes generation of surface difference maps and change identification. The results for each step in the procedure are described below. As the compared data sets are merged into one with different epoch numbers, all the results will be shown in a merged version.

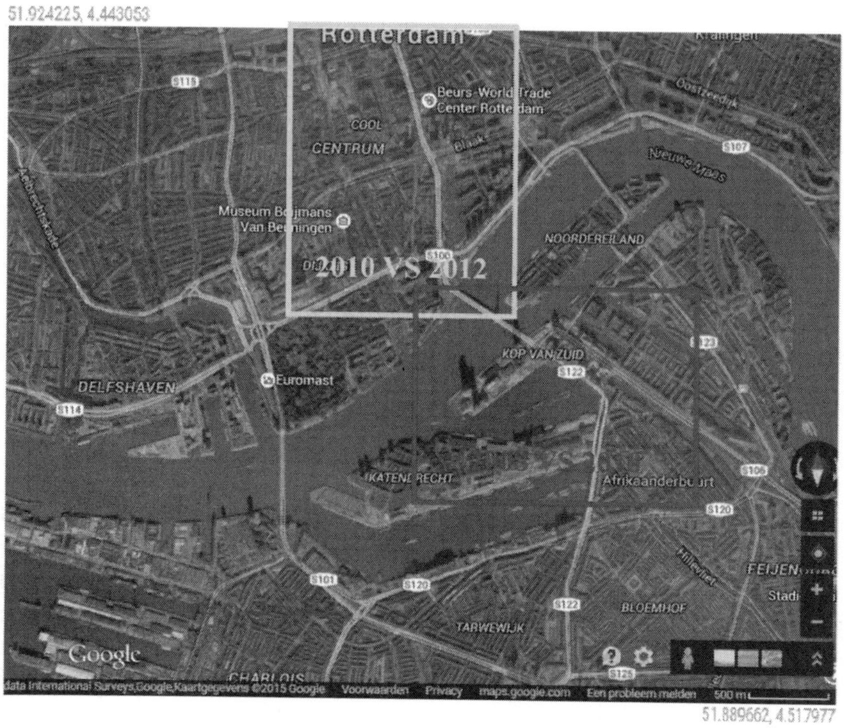

Figure 10. Locations of test areas and their data sets. (From Google earth and the coordinates are approximately measured from the map).

5.1. Detecting Change

Results of the change detection step are shown and the causes of errors in detecting change to buildings are discussed here. The true positives (TP), false positives (FP), and false negatives (FN) are counted "per change" by three people after visual inspection of the compared data sets, the corresponding surface difference map, and our change detection results on a computer screen. To provide some context, an area with 10 m × 10 m around the detected change is shown, together with a high resolution aerial image of the area. The experts indicate whether the detected change is a change to a relevant object. The difference map is also used to verify whether all geometric differences resulted in a detected change. If it is a relevant changed object and we detected it correctly in our result, the change is counted as true positive. If we found a changed object in our result but three experts considered it not a change, the change is a false negative, and if the experts found some changes that we did not detect in the result, the change is a false positive. The completeness and correctness are calculated with the following equations:

$$completeness = \frac{TP}{TP + FN} \tag{1}$$

$$correctness = \frac{TP}{TP + FP} \tag{2}$$

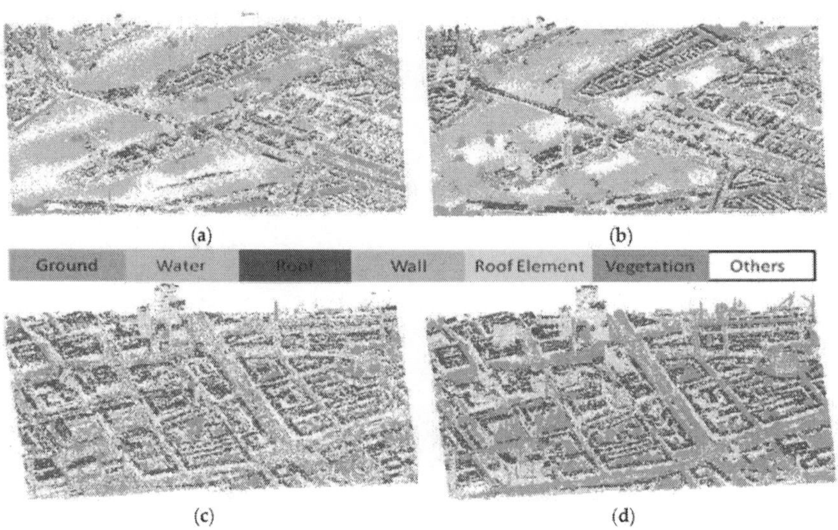

Figure 11. The scene classification results for the test areas. (**a**) 2008—commercial area; (**b**) 2010—commercial area; (**c**) 2010—residential area; (**d**) 2012—residential area.

It is notable that one large building change is counted as one change, the same as a small change, regardless of the number of points within the change. In order to know how the scene classification result affects the change detection result, we also calculated the percentage of errors that are caused by the scene classification result.

5.1.1. Surface Difference Mapping

A surface difference map is generated from original data sets, regardless of the scene classification results. Some difference maps have already been shown in Figure 1 (Section 3.1) and, as discussed, the point density does not affect the surface difference map. More difference maps can be seen in Figure 12. The surface difference map gives only clues as to where changes may have occurred in the data sets. To improve their display, all the data sets have been merged into one. Different colors show different values of the differences between the compared data sets.

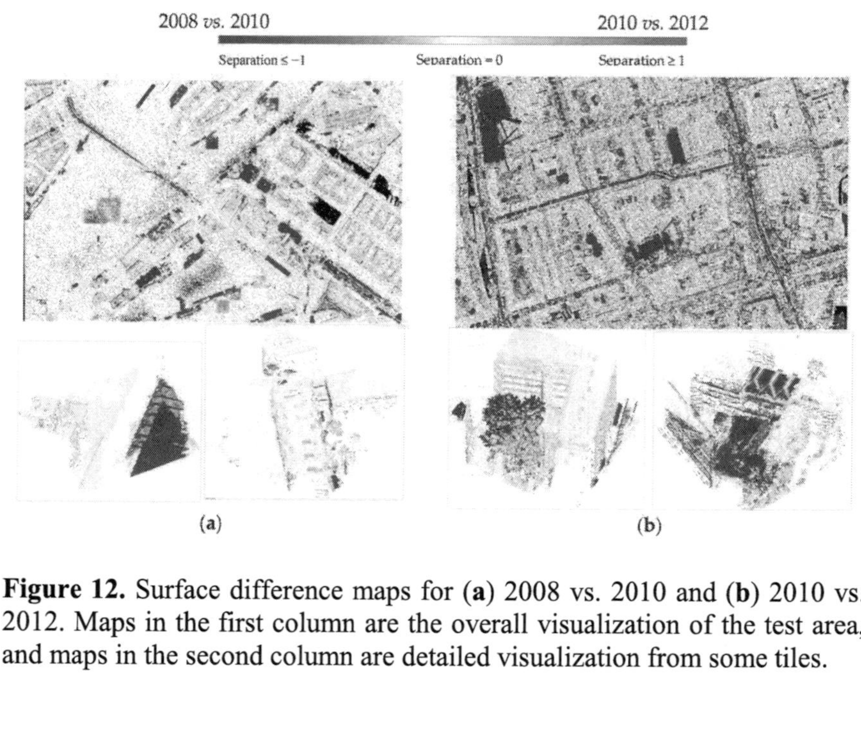

Figure 12. Surface difference maps for (**a**) 2008 vs. 2010 and (**b**) 2010 vs. 2012. Maps in the first column are the overall visualization of the test area, and maps in the second column are detailed visualization from some tiles.

In Figure 12, significant differences can be directly observed on the surface difference maps where the colors become deeper. Red points appearing in the middle of water in Figure 12a indicate large differences in water surfaces. This large difference is caused by lack of points on the water in one of the epochs, while points were recorded in the other epoch. These difference maps are the inputs for the change identification step.

5.1.2. Change Detection Results

Using the surface difference map, we labelled the data sets as "changed", "unchanged", and "unknown". Figure 13 shows some results of both types of change (an entire building change (a), as well as changes to building elements, e.g., changes (b) and (c)). The interpretations "demolished" and "new" can be decided from the sign of the surface difference value in the difference map. For example, the points seen in epoch 2008 but not in epoch 2010, are "demolished" objects. A small change is shown horizontally for clearer visualization in Figure 13c. As there are new points in epoch 2010, as well as demolished points beneath them in epoch 2008, we can infer that there are some extensions of the objects. To understand which objects caused the extension, the changes need to be classified.

We selected 20 tiles with changes randomly from each test area to evaluate the performance of the change detection method. There are more false positives than false negatives, so completeness is greater than accuracy, i.e., 50% (Table 4) of identified changes are not relevant. However, the vast majority of changes to buildings are actually detected, making it convenient to only look at the detected changes and ignore the irrelevant ones.

As shown in Table 5, 40% of the errors (false positives and negatives) are due to incorrect scene classification in Test area 1, and 67% in Test area 2. In Test area 1, which is a commercial area of Rotterdam, the main propagation errors from scene classification are large and long balconies in walls, which are incorrectly classified as roofs. In Test area 2, a residential area, the error propagation mainly arose from trees that are so dense that they were wrongly classified as building roofs. So, changes in trees were incorrectly detected as changed building objects.

Note that in Table 5 there are two false negatives (relevant changes that are not detected) due to our method in Test area 1. These two relevant changes are detected in the change detection method, but the 3D bounding boxes are not generated. The reason for this is that we applied a threshold value for the ratio of the length to width to exclude some thin and long objects such as fences and railings during the change quantification stage. These two relevant changes were roofs that are long and narrow, so 3D bounding boxes were not generated for them.

The advantage of using raw lidar data, instead of using DSM or image, is that lidar data can penetrate trees and the information under trees can be preserved. If a change happens under trees, DSM or image may not detect the change. One such case is shown in figure (i) in Section 5.2.1.

Figure 13. (**a**,**b**) Examples of large changes in buildings; (**c**) changes to dormer sizes in the merged data sets (removed dormers and extensions to dormers in height and length).

Table 4. The completeness and correctness of detected change.

	True Positive	False Positive	False Negative	Correctness	Completeness
2008 *vs.* 2010	34	34	3	50.00%	91.89%
2010 *vs.* 2012	35	30	6	53.86%	83.33%

Table 5. Percentage of errors due to scene classification and errors due to the method of change detection used.

	False Positive		False Negative		
2008 *vs*. 2010 (Test area 1)	34	14	3	1	Errors due to the scene classification error 40%
		20		2	Errors due to limitations of the change detection method 60%
2010 *vs*. 2012 (Test area 2)	30	18	6	6	Errors due to the scene classification error 67%
		12		0	Errors due to limitations of the change detection method 33%

5.1.3. Change Detection Error Due to Scene Classification Error

The points on the surface difference map that are labelled as part of a building are selected for the building change-detection step. When buildings are not correctly classified in the scene classification in any one of the epochs, it will influence the quality of building change detection. Errors in scene classification give rise to false positive and false negative results.

False positives occur when a change has occurred in the areas where non-building objects have been incorrectly classified as “buildings”. In the yellow rectangles shown in Figure 14, there is an error in the scene classification of one of the epochs such that cars are incorrectly classified (in the top row) as a building roof; the water surface is also incorrectly classified as a building roof (in the bottom row). As there are changes in these areas, and one of them is incorrectly classified as part of a building, the changes are confirmed as a “change to a building”. Although these represent real changes that have occurred, they are not changes to buildings. In fact, most false positives occur with trees, when a tree is incorrectly classified as a building roof in one of the epochs.

False negatives occur when a change is confirmed but this has not been classified as being part of a building in the data sets being compared: the change will not be signalled as a “change to a building”. Some examples of false negatives resulting from scene classification errors are shown in Figure 15. If one of the data sets is correctly classified, false negatives can be avoided.

In addition, cranes, bridges, etc., are sometimes classified as being part of a building. Nevertheless, we did not analyze changes in these objects because they did not fall within the scope of our research and we had not defined their features. Consequently, errors related to these objects have been ignored and not discussed in this paper.

5.1.4. Errors in Change Detection Due to Our Algorithm

In addition to errors propagated from scene classification results, other errors arise that are caused by limitations of the change detection algorithm used. From visual inspection, we could conclude that no changes were missed (false negatives) due to our change detection method; some unchanged walls were, however, classified as changed (false positives). Figure 16a shows the incorrect classification of occluded walls as “changed”. We assumed that if no change to the roof of a building had occurred, the attached walls would not change either, even if they had a large difference value. Under this assumption, all wall points in the left-hand image of Figure 16a should be labelled "unchanged". However, there are some projections—balconies, sun shades, etc.—on the walls that are far away from the roof, and these projections will be identified as "changed" because they have a large difference value and there was no unchanged roof found in their 2D neighborhood.

Figure 16b shows an example of what should be "unknown" points being incorrectly labelled as "unchanged". The lack of data for the roofs in one epoch was caused by a water layer that absorbed the laser signals. In the other epoch, the point distribution is rather regular. We expected that the entire area for which data are missing would be classified as "unknown". However, only the central part of the area was classed as "unknown" (light blue). The reason is because only the central parts of these areas in the merged data sets have a large difference value (>1.0 m radius). In other parts there are points from another epoch with a difference value less than 10 cm. Finally, these errors are not important because they did not influence the identification of the "changed" areas, they only resulted in a mix up between the designation "unchanged" and "unknown".

5.2. Change Classification

5.2.1. Results

The change classification results for the two test areas are visualized in Figure 17; for examples with a higher level of detail see Figure 18. We chose several sites where changes were successfully detected and properly classified. These sites included examples of: (a) newly built dormers on roofs; (b) lack of data for roofs in one epoch (unknown) and lack of data for walls and ground because of occlusion; (c) undefined changes on roofs; (d) newly built constructions above roofs; (e) cars parked on top of buildings; (f) newly built and demolished buildings (two examples in one image); (g) add-on building constructions; and (h) insulation layers added to roofs.

Some change detection errors are also caused by errors in scene classification—in addition to those arising from the limitations of our algorithm. These are discussed in Subsection 5.1.2. More false positives are observed, lowering the accuracy, which is analyzed in Section 5.3.

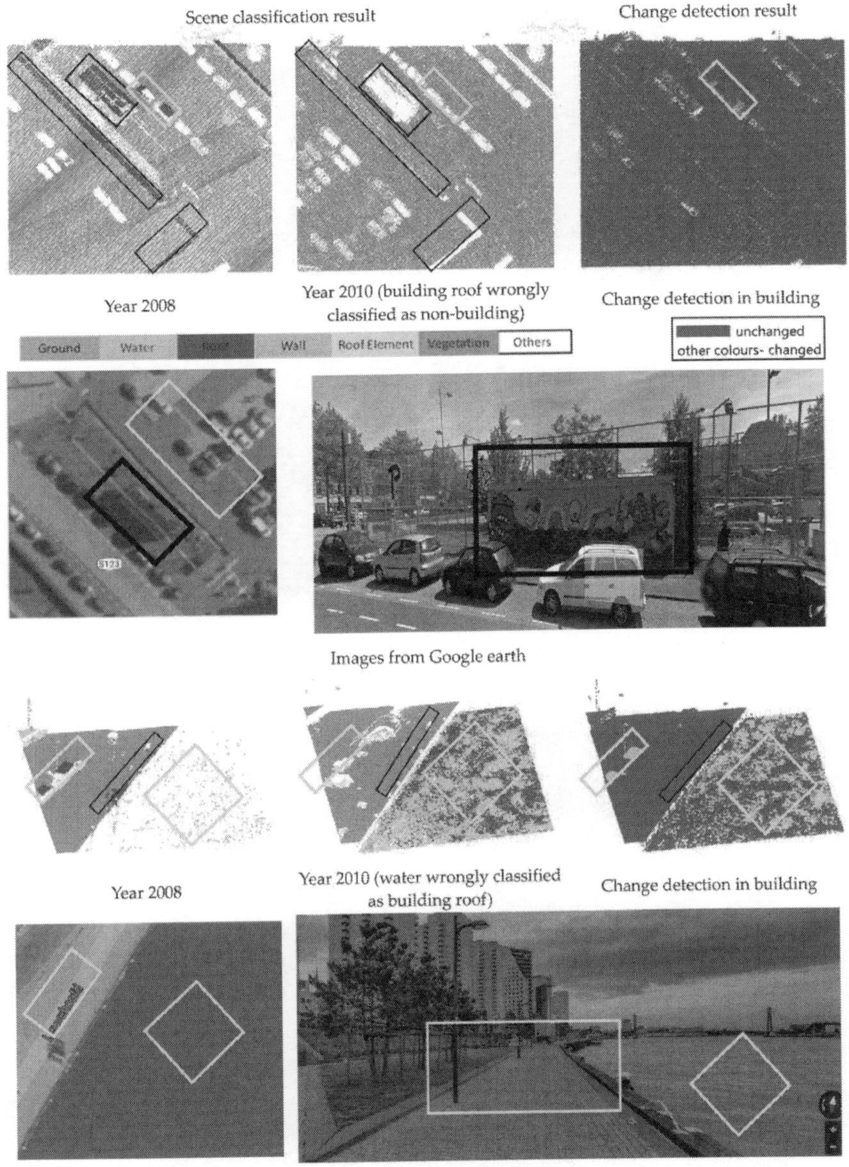

Figure 14. Scene classification errors that have no influence on the change detection results (black rectangles), and ones that have a negative influence (yellow rectangles).

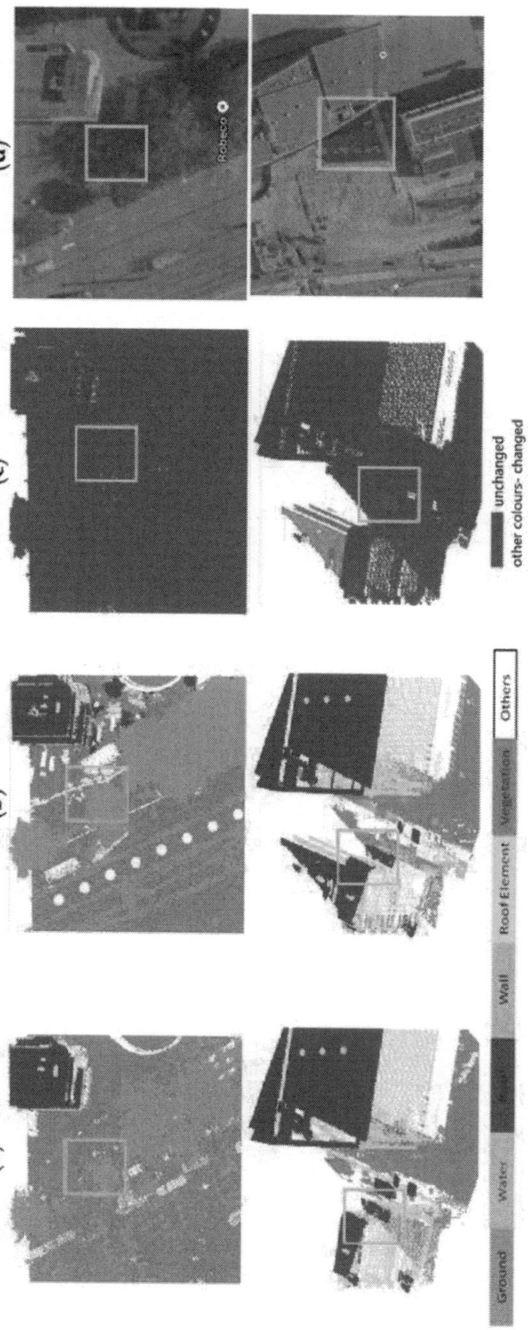

Figure 15. Examples of false negatives caused by scene classification errors. Because a building has been classified as an "undefined object" in the 2012 data set, changes to this building will not be detected (Column 3). (**a**) Data set 2010; (**b**) Data set 2012; (**c**) Change detection in building (merged); (**d**) Image from Google maps.

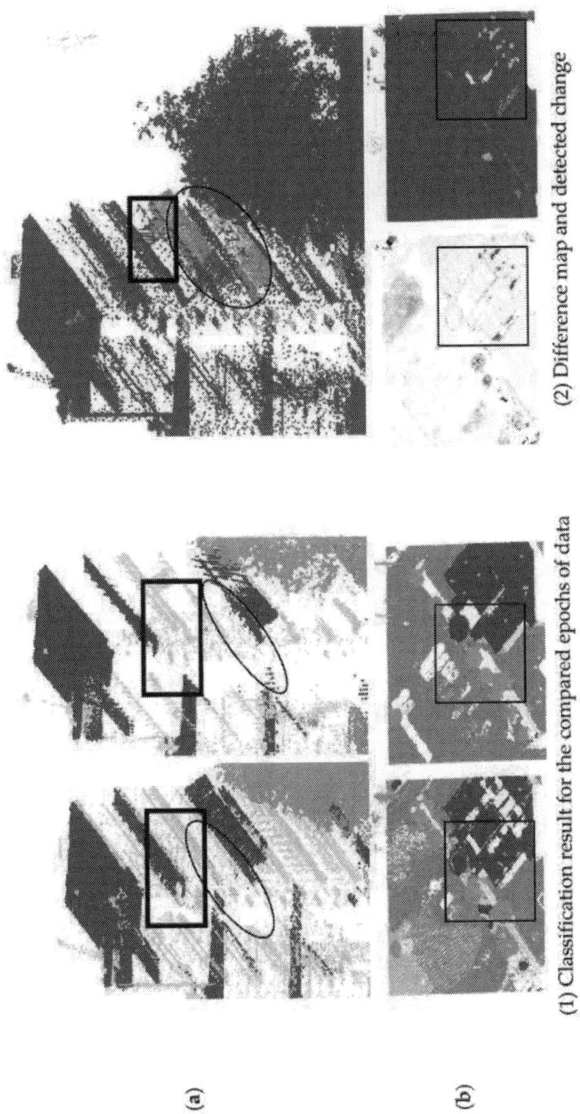

Figure 16. Errors due to our change detection algorithm. Colors in this figure represent the same objects as in Figure 15. In Figure 16, group (1**a**) shows some errors in the ellipse that are due to scene classification errors, and some in the box are due to lack of the contextual information (wall points far away from the roof). Group (1**b**) shows unknown points labelled as unchanged. As seen in the classification result of Group (1b, left-hand), there is a lack of data (some holes) due to limited reflection (red) in the left-hand image of the roof, but the roof in the right-hand image is quite well covered. We expected the entire area where data is lacking to be labelled "unknown", but in the result in (2) only the central parts of the "gaps" are "unknown" (light blue), while other parts are labelled "unchanged" (brown).

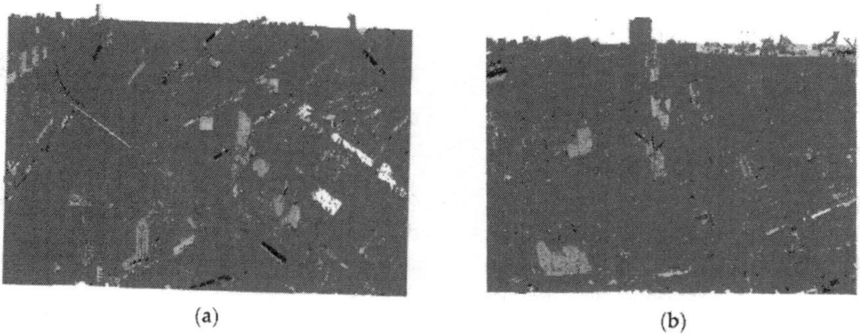

Figure 17. Classification of change in the test areas. (a) 2008 vs. 2010, (b) 2010 vs. 2012.

5.2.2. Error Analysis

We randomly chose another 20 tiles with different changes from each study area to cover all types of changes on roofs. Error analysis of the change classification was done by manually assessing the correctness and completeness of changes detected (see Table 6).

The accuracy of change classification for roofs and walls is high compared to other objects because changes to roofs and walls occur most often in buildings undergoing entire change. In addition to that, scene classification is more reliable for larger objects: changes to buildings are generally large and can easily be correctly classified. Small changes are not so easily separated into their correct change classifications. Most constructions and undefined objects on rooftops are incorrectly classified as new or demolished dormers, especially if they are near a pitched roof. We have assumed that dormers are normally located near the pitched roof, but often roofs are in fact a combination of flat and pitched roofs, and there are as many changes on these roofs, which are not changed dormers but incorrectly classified as dormers. It is also difficult, based only on their size, to distinguish large cars on a rooftop parking lot from building constructions. As a result, some of these large cars are classified as constructions. We conclude that, even if the changes are detected correctly, in highly complex urban areas it is hard to completely and correctly classify the detected changes using geometrical and relational rules.

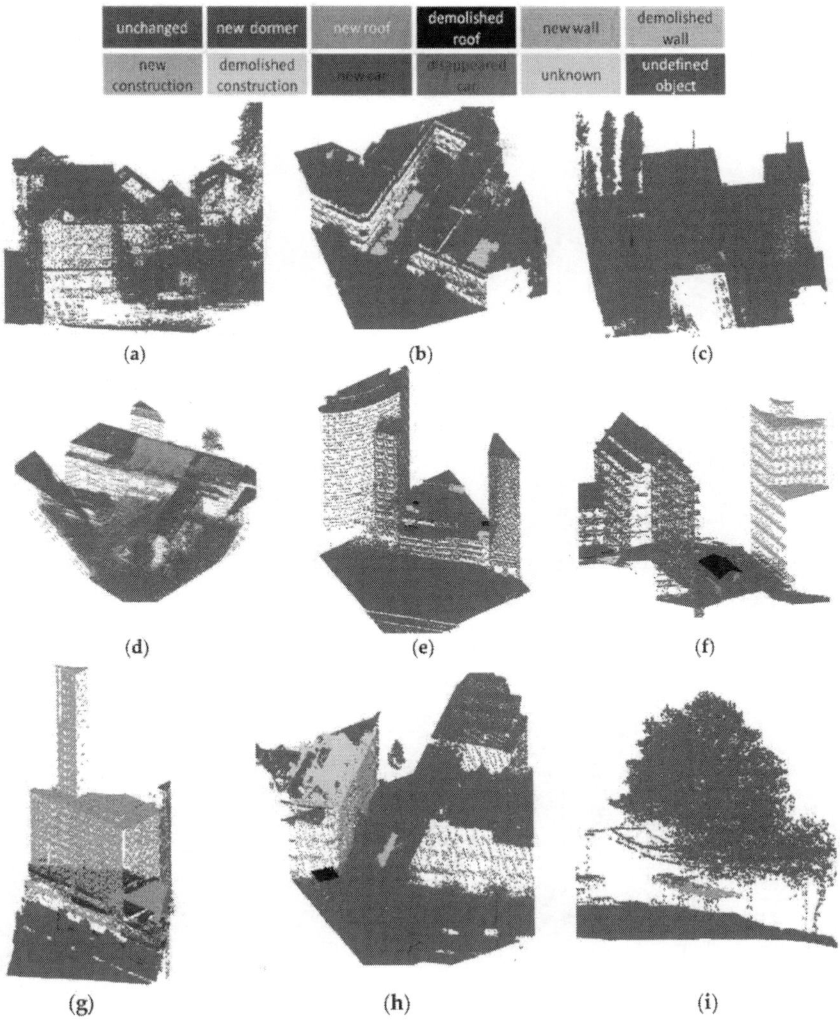

Figure 18. Successfully detected and classified examples of change. (**a**) newly built dormers on roofs, (**b**) lack of data for roofs in one epoch (unknown) and lack of data for walls and ground because of occlusion, (**c**) undefined changes on roofs, (**d**) newly built constructions above roofs, (**e**) cars parked on top of buildings, (**f**) newly built and demolished buildings (two examples in one image), (**g**) add-on building constructions, (**h**) insulation layers added to roofs. (**i**) new building roof under trees.

Table 6. Completeness and correctness of the changed objects.

Label	2008 *vs.* 2010			2010 *vs.* 2012		
	Correctness	Completeness	Number of Objects	Correctness	Completeness	Number of Objects
Undefined object	91%	35%	29	100%	26%	23
New dormer	6%	100%	1	80%	24%	17
Demolished dormer	-	-	0	-	-	0
New car on roof top	33%	11%	9	50%	7%	14
Car no longer on rooftop	100%	100%	1	-	-	0
New construction on rooftop	13%	25%	8	42%	56%	9
Demolished construction on rooftop	100%	40%	5	-	0%	2
New roof	78%	81%	26	59%	100%	17
Demolished roof	91%	100%	20	90%	100%	18
New wall	100%	100%	8	100%	100%	7
Demolished wall	100%	100%	8	100%	100%	4

5.3. Object-Based Analysis

5.3.1. Results

Minimum 3D bounding boxes are generated around the connected components to enable selection of relevant changes in the buildings; for some results see Figure 19. The location of the center point, the area, and the volume of the minimum 3D bounding box are calculated. Changes are shown in Figure 19a. Figure 19b,c give two examples of 3D bounding boxes: Figure 19b shows a building object that has undergone complete change; and Figure 19c shows a changed building element (newly built dormer). The bounding box in Figure 19c is larger than the dormer because of the influence of some outliers occurring in the same plane as the dormer.

5.3.2. Error Analysis

We used connected component labelling to form objects and their 3D bounding boxes to estimate their area and volumes. We found that connected component labelling failed to correctly form objects under two circumstances:

(1) False positives in buildings

Sometimes points on walls or roofs are false positives. These points may belong to an unchanged wall that has been incorrectly detected as changed (see discussion in Subsection 5.1.4), or they can be sparse points of plants on a balcony or rooftop (irrelevant changes). If such points are close together, connected component labelling will form an object that is as big as an entire wall or even a complete building; see Figure 20a.

(2) Small objects that are too close to each other

Small objects that are too close to each other will be connected together as one object. This often occurs with cars parked on rooftops of buildings and shelters at bus stops that are very close to each other. Figure 20b shows an example of cars parked on a rooftop. The consequence is an incorrect shape of detected change.

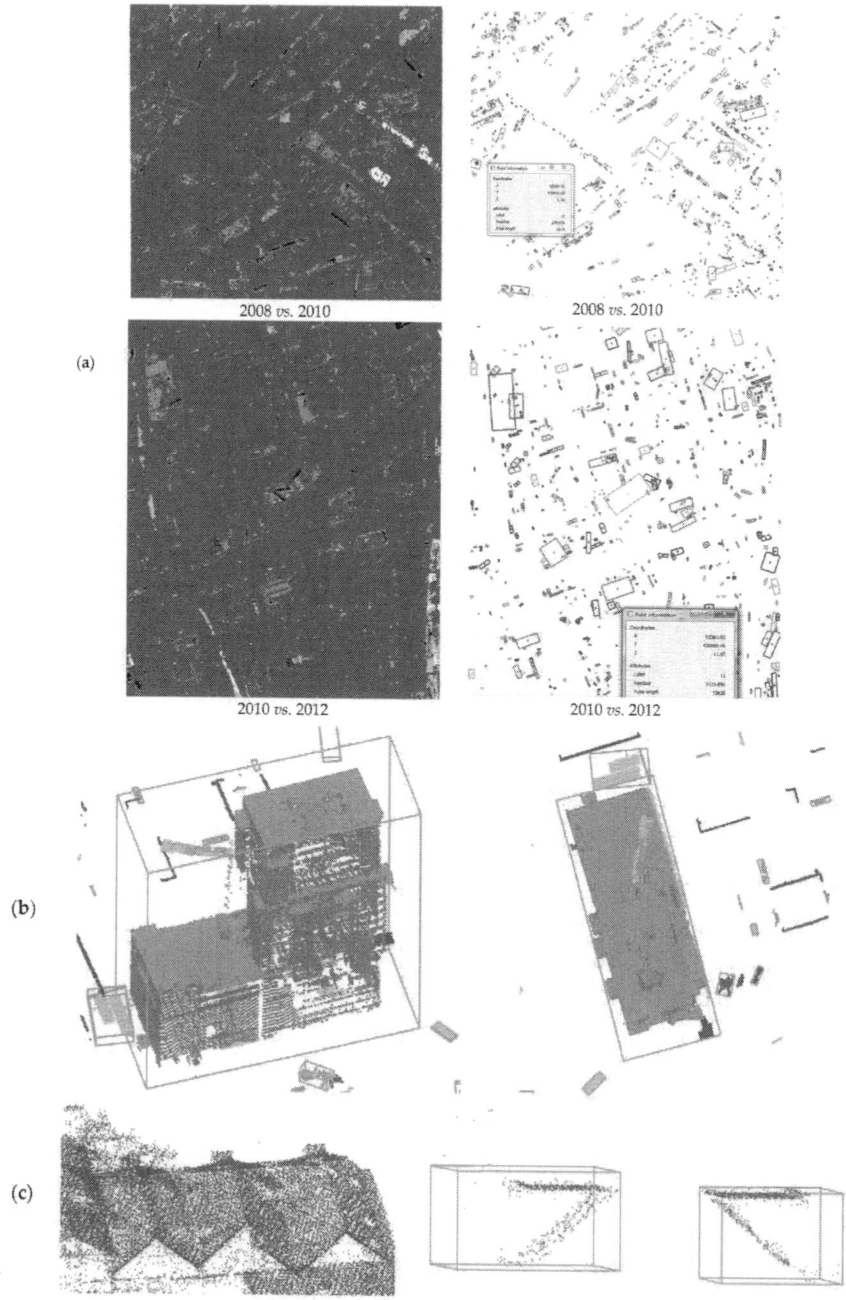

Figure 19. 3D bounding boxes and the center points of the relevant changes, together with the calculated area and volume; the label, area and the volume are shown in the point number window. In this figure, points without bounding boxes are irrelevant changes. (**a**) 3D bounding boxes for the two test areas; (**b**) an example of 3D bounding box for an entire new building; (**c**) an example of 3D bounding box for newly built dormers.

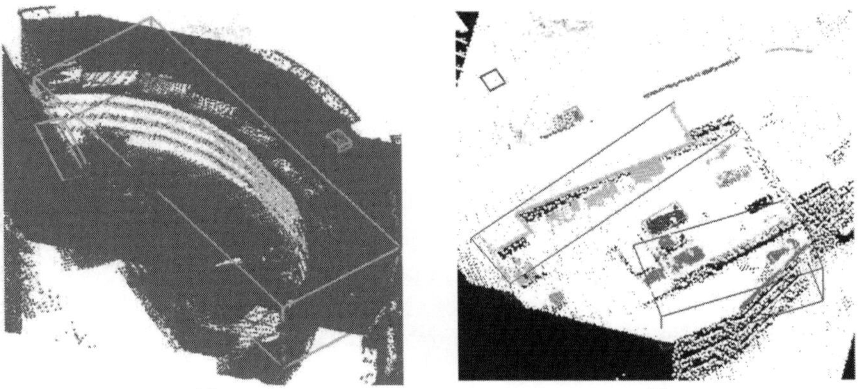

(a) (b)

Figure 20. 3D bounding boxes of some wrong changes and irrelevant changes. (**a**) Incorrect changes in walls and irrelevant changes on roofs; (**b**) Cars that are close to each other and close to some fences are grouped to large objects. Yellow cubes are show objects classified as new buildings. Blue cubes are new unknown objects. Green cubes are new constructions on top of the roof.

6. CONCLUSIONS

In this paper, we present a method for detecting and classifying changes to buildings by using classified and well registered (strip difference <10 cm) laser data from several epochs. The analysis of our results leads us to draw several conclusions.

Provided distances between surfaces are greater than 10 cm and the area of change more than 4 m^2, both large and small changes can be automatically detected using a surface difference map and a rule-based change detection algorithm. Areas classified as "unknown" can be correctly identified in cases of occlusions and water reflection. The surface difference map is not affected by the point density of the compared data sets. Some of the other parameters, such as the number of points indicating the approximate area and radius for the surface growing method, should be adjusted with varying point densities.

Larger changes can be correctly assessed as belonging to a building provided that building has been correctly classified as such in the scene classification step for one of the data sets being compared. The accuracy of object recognition was evaluated by overlaying the 3D bounding boxes of the buildings for which change was detected on manually generated reference data. Our method detected 91% of actual changes in Test area 1 and 83% in Test area 2. Nearly half of the changes detected in objects were irrelevant changes.

About half of the false positives that occurred were caused by scene classification errors. The other false positives can be mostly attributed to our change detection algorithm. Mostly they are spurious changes in a wall that is at a larger distance to a roof, plants growing in a rooftop garden, containers, isolation layers, cleaning machines on the top of buildings, etc. Regarding classifying changes, we conclude that ,even if changes are identified correctly, in highly complex urban areas it is difficult to completely and accurately classify the smaller detected changes using geometrical and relational rules.

The classification result for the changed object shows that large changes affecting an entire building, for example its roof or a main wall, can be detected with a higher degree of accuracy than changes made to a building element, such as dormers or construction on top of the building.

Overall, our method detected 80-90% of changes to buildings. It is the first time that we tried to detect high-level detail changes in buildings using raw lidar data, so we did not compare our accuracy to other methods. Our method is, however, not yet capable of distinguishing small irrelevant changes to objects from relevant ones. A more accurate definition of a "changed building object" and its characteristics are required in order to better interpret differences between two data sets.

ACKNOWLEDGMENTS

The authors would like to thank the Municipality of Rotterdam for providing the Lidar data sets.

AUTHOR CONTRIBUTIONS

Sudan Xu designed the algorithms as described in this paper during her PhD research, and was responsible for the main organization and writing of the paper. Sander Oude Elberink and George Vosselman supervised Sudan during her PhD project contributed to paper writing.

CONFLICTS OF INTEREST

The authors declare no conflict of interest.

REFERENCES

1. Xiao, W.; Vallet, B.; Paparoditis, N. Change detection in 3D point clouds acquired by mobile mapping system. ISPRS Ann. Photogramm. Remote Sens. Spat. Inf. Sci. **2013**, 2, 331–336.
2. Xu, S.; Oude Elberink, S.; Vosselman, G. Multiple-entity based classification of airborne laser scanning data in urban areas. ISPRS J. Photogramm. Remote Sens. **2013**, 88, 1–15.
3. Mas, J.-F. Monitoring land-cover changes: A comparison of change detection techniques. Int. J. Remote Sens. **1999**, 20, 139–152.
4. Byrne, G.F.; Crapper, P.F.; Mayo, K.K. Monitoring land-Cover change by principle component analysis of multitemporal landsat data. Remote Sens. Environ. **1980**, 10, 175–184.
5. Sharma, B.; Rishabh, I.; Rakshit, S. Unsupervised change detection using RANSAC. In Proceedings of the IET International Conference on Visual Information Engineering, Bangalore, India, 26–28 September 2006; pp. 24–28.
6. Yang, Z.; Qin, Q.; Zhang, Q. Change Detection in high spatial resolution images based on support vector machine. In Proceedings of the IEEE International Geoscience and Remote Sensing Symposium, Denver, CO, USA, 31 July–4 August 2006; pp. 225–228.
7. Di, F.; Li, X.; Zhu, C. A new method in change detection of remote sensing image. In Proceedings of the 2nd International Congress IEEExplore Digital Library, Image and Signal Processing (CISP' 09), Tianjin, China, 17–19 October 2009; pp. 1–4.
8. Tanathong, S.; Rudahl, K.T.; Goldin, S.E. Object oriented change detection of buildings after a disaster. In Proceedings of the ASPRS 2009 Annual Conference, Baltimore, MD, USA, 9–13 March 2009. On CD-Rom.

9. Kasetkasem, T.; Varshney, P.K. An Image Change detection algorithm based on Markov Random Field models. IEEE Trans. Geosci. Remote Sens. **2002**, 40, 1815–1823.

10. Murakami, H.; Nakagawa, K.; Hasegawa, H.; Shibata, T.; Iwanami, E. Change detection of buildings using an airborne laser scanner. ISPRS J. Photogramm. Remote Sens. **1999**, 54, 148–152.

11. Khoshelham, K.; Oude Elberink, S.J.; Xu, S. Segment based classification of damaged building roofs in aerial laser scanning data. IEEE Geosci. Remote Sens. Lett. **2013**, 10, 1258–1262.

12. Gerke, M. Dense matching in high resolution oblique airborne images. Int. Arch. Photogramm. Remote Sens. Spat. Inf. Sci. **2009**, 38, 77–82.

13. Vosselman, G.; Gorte, B.G.H.; Sithole, G. Change detection for updating medium scale maps using laser altimetry. Int. Arch. Photogramm. Remote Sens. Spat. Inf. Sci. **2004**, 34, 207–212.

14. Vögtle, T.; Steinle, E. Detection and recognition of changes in building geometry derived from multi-temporal laser scanning data. Int. Arch. Photogramm. Remote Sens. Spat. Inf. Sci. **2004**, 34, 428–433.

15. Rutzinger, M.; Ruf, B.; Hofle, B.; Vetter, M. Change detection of building footprints from airborne laser scanning acquired in short time intervals. Int. Arch. Photogramm. Remote Sens. **2010**, 38, 475–480.

16. Choi, K.; Lee, I.; Kim, S. A feature based approach to automatic change detection from Lidar data in urban areas. Int. Arch. Photogramm. Remote Sens. Spat. Inf. Sci. **2009**, 38, 259–264.

17. Matikainen, L.; Hyyppä, J.; Kaartinen, H. Automatic detection of changes from laser scanner and aerial image data for updating building maps. Int. Arch. Photogramm. Remote Sens. Spat. Inf. Sci. **2004**, 35, 434–439.

18. Rottensteiner, F. Building change detection from digital surface models and multi-spectral images. Int. Arch. Photogramm. Remote Sens. Spat. Inf. Sci. **2007**, 36, 145–150.

19. Champion, N.; Rottensteiner, F.; Matikainen, L.; Liang, X.; Hyyppä, J.; Olsen, B.P. A test of automatic building change detection approaches. Int. Arch. Photogramm. Remote Sens. Spat. Inf. Sci. **2009**, 38, 145–150.

20. Chen, L.; Lin, L.; Cheng, H.; Lee, S. Change detection of building models from aerial images and Lidar data. Int. Arch. Photogramm. Remote Sens. Spat. Inf. Sci. **2010**, 38, 121–126.

21. Awrangjeb, M.; Fraser, C.S.; Lu, G. Building change detection from lidar point cloud data based on connected component analysis. ISPRS Ann. Photogramm. Remote Sens. **2015**, 2, 393–400.

22. Hebel, M.; Arens, M.; Stilla, U. Change detection in urban areas by object-based analysis and on-the-fly comparison of multi-view ALS data. ISPRS J. Photogramm. Remote Sens. **2013**, 86, 52–64.

23. Vosselman, G. Automated planimetric quality control in high accuracy airborne laser scanning surveys. ISPRS J. Photogramm. Remote Sens. **2012**, 74, 90–100.

24. Vosselman, G.; Gorte, B.G.H.; Sithole, G.; Rabani, T. Recognising structure in laser scanner point clouds. ISPRS Arch. Photogramm. Remote Sens. Spat. Inf. Sci. **2004**, 40, 33–38.

25. Vosselman, G. Point cloud segmentation for urban scene classification. ISPRS Arch. Photogramm. Remote Sens. Spat. Inf. Sci. **2013**, 40, 257–262.

CHAPTER 4

Automatic Object Extraction from Electrical Substation Point Clouds

Mostafa Arastounia * and Derek D. Lichti

Department of Geomatics Engineering, University of Calgary,
2500 University Drive NW, Calgary, AB T2N 1N4, Canada

ABSTRACT

The reliability of power delivery can be profoundly improved by preventing wildlife-related power outages. This can be achieved by insulating electrical substation components with non-conductive covers. The manufacture of custom-built covers requires as-built models of the salient components. This study presents new, automated methodology to recognize key components of electrical substations from 3D LiDAR data acquired using terrestrial laser scanning. The proposed methodology includes six novel algorithms to recognize key components (fence, cables, circuit breakers, bushings and bus pipes) of electrical substations. Three datasets with different resolutions and configurations are used in this study. A Leica HDS 6100 laser scanner was used to acquire the first dataset and a Faro Focus^{3D} laser scanner was employed to collect the second and third datasets. The obtained results indicate that 178 and 171 out of 181 electrical substation elements were successfully recognized in the first and second dataset, respectively, and 183 out of 191 components were identified in the third dataset. The results also demonstrate that an average 97.8% accuracy and average 98.8% precision at the point cloud level can be achieved.

KEYWORDS

LiDAR; laser scanning; object recognition; segmentation; electrical substation

1. INTRODUCTION

Modern power distribution systems provide high service reliability. According to the US Department of Energy, today's electricity system is 97.97% reliable [1]. However, power outages and interruptions that can last for minutes to hours still occur. Power outages can be extremely expensive, costing the US Government $150 billion annually [1]. Studies on 203 power cuts in the US revealed that the repair costs range from $500 to more than $140,000, comprising equipment replacement and repair [2]. In addition to economic loss, power cuts have a negative impact on system reliability and customer satisfaction [3]. Wildlife is the third leading-cause of power outages, after natural disasters and human-caused incidents, since electrical substations are constructed in open areas and can be easily reached by wildlife [3]. This study is aimed at developing a solution to aid in the prevention of wildlife-related power outages.

Many different solutions have been proposed to prevent power outages. Burnham et al. [4] propose the use of cover-up materials and also suggest some modifications to the existing standard design of electrical substations such as pole-line reconfiguration. Sundararajan et al. [5] propose using perch guards, metal or plastic spikes and insulating caps to reduce bird-related outages. They also suggest constructing alternative perch sites close to substations to prevent birds from perching on transmission lines, in substations and on over-head distribution lines. Harness [6] investigates which species come into contact with electrical equipment in different regions, and proposes covers, guards and reconfigurations specific to species causing power outages in a given region. The studies in [4,5,6] do not address the issue with a comprehensive solution and they only focus on the prevention of bird-related power outages by proposing manual approaches. By applying modifications and insulations proposed in [4,5,6], other sorts of wildlife, such as reptiles and mammals, can still cause power cuts.

Covering the conductive elements of electrical substations with insulating materials can be a comprehensive solution to power outages caused by wildlife. Since the configuration of electrical substations can substantially differ from one to the next, non-conductive covers need to be custom-manufactured. The manufacture of the covers requires both an accurate inventory of substation components and accurate (centimeter-level) dimensions. In addition to site-to-site differences, the configuration and design of any given substations can change over time. Thus, any existing, as-designed plan is unlikely to be an accurate record of the current state of a site. Manual measurement of electrical equipment dimensions is not possible since safety restrictions require the substation to be de-energized, which is extremely costly. Technologies, such as laser scanning, offer safe and non-contact means of rapidly collecting substation data. Millions of three-dimensional (3D) data points can be measured within minutes. The key to exploiting these data for accurate dimensional characterization of substations is the automated recognition of relevant components from the data, which is the subject of this work.

2. LITERATURE REVIEW

To the best of authors' knowledge, no comprehensive automated solution has been proposed to recognize all components of electrical substations from LiDAR data. Therefore, a review of studies that propose specific approaches for substation documentation using optical methods is first provided. This is followed by a review of methods for the segmentation and object recognition from LiDAR data in environments that share similarities with electrical substations. The similarities are in terms of the types of geometric primitives found on substation sites, i.e., linear, planar and cylindrical surfaces.

Although other studies have investigated the use of optical methods for substation documentation, they do so with manual methods or limit their scope to the recognition of specific components. Sternberg and Kersten [7] manually construct a 3D model of an electrical substation by using commercial-available software. They approximate objects by simple geometric primitives, i.e., electric power cables and insulators are represented by polylines and cylinders, respectively. Xu et al. [8] employ LiDAR data to study the current state of the electrical assets, such as power lines and electrical substations. However, they do not propose a systematic method for data processing and the inspection of electrical assets is carried out visually. Gonzalez-Aguilera et al. [9] employ digital camera images and photogrammetric techniques to model objects in electrical substations. Armeshi and Habib [10] also recognize insulators of a substation site from digital images employing photogrammetric techniques. Photogrammetry can be restrictive in terms of the number of images that must be captured to completely cover a site and the close proximity to the electrical equipment at which they must be acquired, which may endanger the individual taking the images. Arastounia and Lichti [11] recognize insulators from LiDAR data of a subset of an electrical substation point cloud. Both studies in [10,11] limit their scope to insulator recognition and other objects are not recognized. In this contribution, LiDAR data are collected by non-contact terrestrial laser scanners, which are able to scan the site from fewer instrument locations and from a farther distance compared to digital cameras. This significantly lowers the safety risks and duration of data collection. Noting that manual methods are used in [7,8], the aim of our contribution is to develop methods for automated recognition of electrical substations key components so that dimensions can be extracted in support of insulating cover manufacture.

Vosselman et al. [12] describe plane and cylinder recognition both by spatial-domain (surface growing) and parametric-domain (3D Hough transform that is an extension of Hough transform algorithm [13]) algorithms. They recommend using spatial-domain methods for processing datasets larger than twenty million points. Belton and Lichti [14] use local covariance analysis to segment smooth surfaces, edges and boundaries in an industrial site comprising planar and cylindrical surfaces. However, industrial objects are not individually recognized since they primarily focus on the segmentation. Rabbani et al. [15] propose a segmentation algorithm that uses a smoothness constraint to segment LiDAR data of industrial sites incorporating planar and cylindrical surfaces. The focus of this work is, however, limited to the segmentation of smoothly connected areas and objects in the site are not individually recognized. Pu and Vosselman [16,17,18] extract building planar features, such as walls, roofs, and doors. First, planar surfaces are segmented and then some constraints, such as size, position, orientation, and point density are applied to recognize various features. Planar surfaces are segmented using a surface-growing algorithm, as explained in [12]. Lehtomaki et al. [19] extract poles in a road environment using structured point clouds (scan lines). A scan line segmentation method is carried out based on proximity of points and the resulting segments are further analyzed to detect poles based on their size, shape and orientation. Martinez et al. [20] extract building facades in which planar surfaces are identified by RANSAC (RANdom Sample Consensus) [21]. Arastounia and Lichti [22] propose a new algorithm for plane segmentation, based on calculating the extent of points' neighborhood in three cardinal directions and then compare it to principal components analysis (PCA). Although a LiDAR dataset of an electrical substation is used in this research, they only focus on the planar surface segmentation and no electrical equipment is recognized. Zhu et al. [23] reconstruct building 3D models using airborne laser scanner (ALS) data and topographic data in which planar surfaces (roof patches) are segmented by using PCA and eigenvalues analysis.

Much recent work has also focused on segmentation and recognition of linear features from LiDAR data. Melzer and Brieser [24] introduce an algorithm for extracting power lines from ALS data in which the 2D Hough transform and RANSAC are employed to detect power line segments. This study assumes that power lines are parallel, which is not always the case. Jwa et al. [25] propose an algorithm to reconstruct power cables from a voxel-based, piece-wise line detector (VPLD). The entire dataset is divided into 3D voxels and both eigenvalue analysis and Hough transform are employed to detect linear neighborhoods. The results suffer from under- and over-segmentation, which are refined in [26] by applying topological relationships among power cables. Sohn et al. [27] recognize power lines from ALS data using the method proposed in [25]. However, they employ Markov Random Field (MRF) classifier to identify power lines from the linear segments based on segment height, slope and parallelism. Zhang et al. [28] classifies urban features such as ground, buildings and power lines from ALS data using support vector machine-based classification method. Zhu and Hyyppa [29] extract power lines from ALS data. First, power line candidate points are identified by statistical analysis considering height, point density and histogram thresholds. Candidate points are then converted to 2D images and are further analyzed by image processing techniques. Zhu and Hyyppa [30] model power lines in railroad environment from integrated ALS and MLS data. First, candidate power line points are identified based on some similarity measures, i.e., points' Z gradient with respect to their planimetric co-ordinates. Then, candidate points are converted to 2D images and image-processing techniques are employed to extract power lines. Cables appear as linear objects in both railroad corridors and electrical substation sites. However, railroad cables are usually much longer and their curvature is much less than that of electrical substations cables.

In our contribution, a new, fully-automated methodology is introduced to recognize key components of electrical substations. The proposed methodology is a data-driven approach that exploits unstructured LiDAR data containing only geometric information, i.e., only 3D co-ordinates of points. Furthermore, to improve the data processing efficiency, the data is processed in its original 3D point cloud format without converting it to 2D images or using image-processing techniques.

3. BACKGROUND KNOWLEDGE

This section introduces the key elements of electrical substations and describes their physical characteristics. The key components of electrical substations are: the site boundary fence; cables; circuit breakers; bushings; and bus pipes, as indicated in Figure 1.

3.1. Site Boundary Fence

The fence is an important object since it defines the areal extents of an electrical substation site. It is essentially a set of long, semi-transmissive planar surfaces that are separated by vertical poles. It has a chain-link structure topped with barbed wire. The mesh nature of such a fence makes it a penetrable surface so many laser scanner measurements of the surrounding environment might be collected.

Figure 1. Components of an electrical substation site.

3.2. Cables

Transmission cables within a substation appear as curvilinear objects in LiDAR data. They are essentially space curves with thickness that is small relative to their arc length. They are used to transmit electricity among bus pipes, insulators, and bushings.

3.3. Circuit Breakers, Bushings and Insulators

Circuit breakers are characterized by a set of adjacent, bounded planes that may have two general physical shapes. A regular circuit breaker appears as a box-shaped object consisting of four planar surfaces. An oil circuit breaker comprises three planar surfaces attached to a cylindrical object containing the oil (the fourth plane is occluded). Both types have bushings on top. Insulators are similar to bushings in appearance, though bushings lie above circuit breakers while insulators are located beneath distribution busses or transmission busses or hang from a cross arm on a tower.

3.4. Bus Pipes

Bus pipes are cylindrical-shaped objects that are used in transmission busses, distribution busses and complex steel structures to interconnect electrical equipment. They are primarily utilized where high strength and low conductivity are required [31]. Table 1 summarizes the physical shape of the electrical substations key components that was utilized for their recognition.

Table 1. Physical shape of electrical substations key components.

Object Type	Physical Shape
Fence	Planar
Cable	Curvilinear
Circuit breaker	Planar
Bushing	Cylindrical
Bus pipe	Cylindrical

4. METHODOLOGY

The methodology depicted in Figure 2 is described in the following five subsections. In Section 4.1, points on the ground are extracted and removed. Then, points belonging to the fence and cables are identified in the following two sections. The circuit breakers and bushings are recognized in Section 4.4, which is followed by Section 4.5 where points belonging to the bus pipes are recognized.

In many parts of the proposed methodology, local neighborhood statistics derived from PCA are used as cues. The basis of PCA is the construction and eigenvalue decomposition of the covariance matrix of 3D co-ordinates within each point's neighborhood, i.e.,

$$AV = \lambda V \tag{1}$$

where: A: square 3 × 3 covariance matrix of 3D co-ordinates within a point's neighborhood; λ: scalar eigenvalue; V: 3 × 1 matrix representing an eigenvector

$$V = \begin{pmatrix} V_x \\ V_y \\ V_z \end{pmatrix}_{3\times 1} \Rightarrow \vec{v} = (V_x, V_y, V_z) \tag{2}$$

Considering the dimensions of the covariance matrix (3 × 3), Equation (1) leads to three scalar eigenvalues (λ_1, λ_2, λ_3 assuming $\lambda_1 \geq \lambda_2 \geq \lambda_3$) and three eigenvectors ($\vec{v}1, \vec{v}2, \vec{v}3$). Each eigenvalue ($\lambda_i$) represents the dispersion of a neighborhood in the direction of its corresponding eigenvector ($\vec{v}i$). The eigenvalues and eigenvectors of the covariance matrix are used to infer local neighborhood structure.

4.1. Ground Removal

Although the ground within a substation site is not an important object, it needs to be removed since it usually constitutes a large proportion of total data volume. Substation sites typically have little topographical variation since they must be constructed on level ground. First, the height of the ground is determined from the height histogram of the entire point cloud. The dominant peak corresponding to points on the ground is automatically identified from which the ground height is determined. Points on or below the ground height are then easily discarded and the remaining points are retained for subsequent processing. Figure 3 shows the height histogram of a sample dataset in which the upper arrow points to the dominant peak representing points on the ground and the lower arrow points to the determined height of the ground. The bin width must large enough to reflect the true local point density. In this study, a bin width of 0.2 m was chosen based on topographical variation in substation sites and the point density of the datasets.

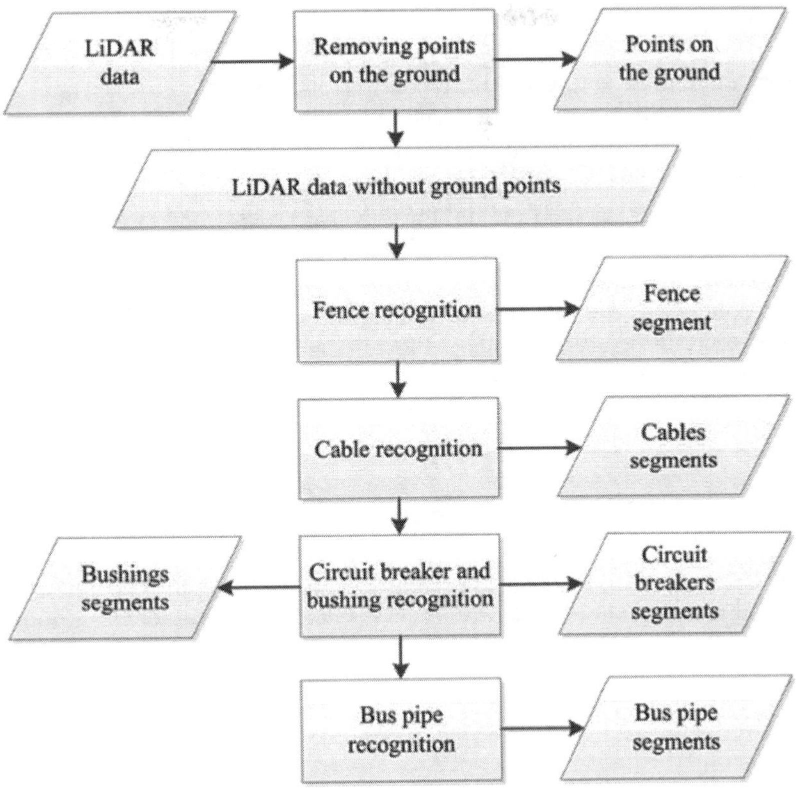

Figure 2. Flowchart of the entire methodology.

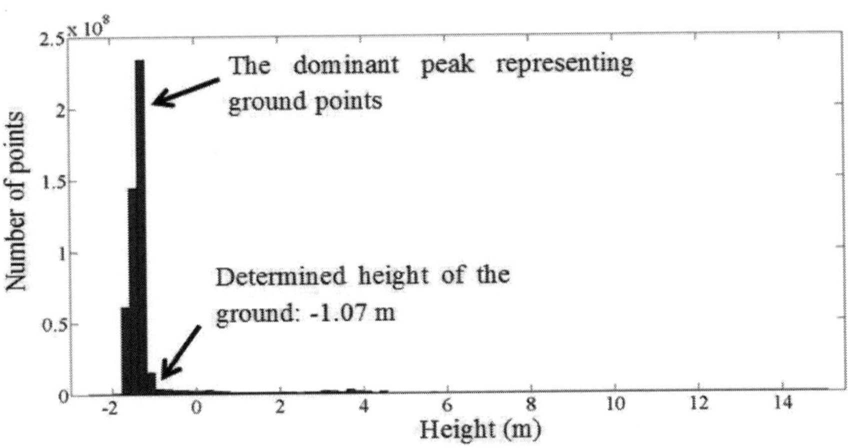

Figure 3. Height histogram of a sample dataset.

4.2. Fence Recognition

The chain link structure of the fence makes it a penetrable surface. As a result, some planar parts of the fence may be sparsely sampled. Thus, two algorithms were developed to recognize the fence. The first algorithm recognizes the fence in datasets in which fence panels are well-sampled, which is the case for the first and third datasets, described in Section 5. The second algorithm is able to recognize the fence without taking the fence point sampling into consideration but is computationally more expensive. The second algorithm is used for the second dataset since the fence in this dataset was sparsely sampled. Flowcharts of both fence recognition algorithms are shown in Figure 4.

4.2.1. First Algorithm for Fence Recognition

First, three-dimensional PCA is employed to detect points belonging to a planar local neighborhood. To detect planar neighborhoods, following conditions need to be met:

$$n\lambda_1 = n\lambda_2 \ and \ n\lambda_3 = 0 \tag{3}$$

where $n\lambda_i$ is a normalized eigenvalue computed as follows:

$$n\lambda_i = \frac{\lambda_i}{\sum_{j=1}^{3} \lambda_j} \tag{4}$$

The neighborhood size used for inspecting the neighborhood structures and the distance threshold used for segmentation in almost all steps of the proposed methodology in this work was 0.15 m. Where a different value is used in the methodology, its rationale is also explained. This threshold (0.15 m) was chosen by considering the size, sparseness and configuration of objects in electrical substation sites. It was found to provide a reasonable balance between incorporating a sufficient number of neighboring points that provide reliable information about local structure and excluding points on adjacent objects.

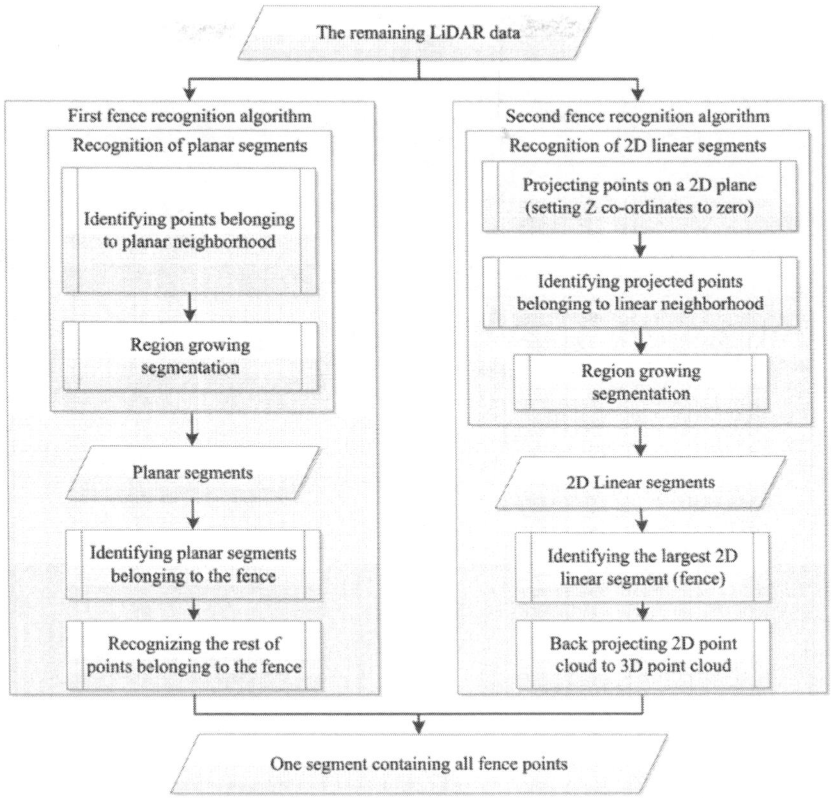

Figure 4. Flowchart of two developed algorithms for fence recognition.

Points within a 3D Euclidean distance of 0.15 m of one another that satisfy the following two conditions are grouped together in a cluster:

1) Lying on a planar surface (meeting the condition in Equation (3))

2) Parallel normal vectors of the best-fit plane

Vertical poles between planar parts of the fence prevent fence points from being clustered into one single segment, as indicated in Figure 5. Hence, the planar fence segments need to be discriminated from planar segments belonging to other objects. To this end, the following two conditions (see Figure 5) must be satisfied for at least four adjacent planar segments (since no other object has this many adjacent planar segments): the normal vectors of adjacent planar segments should be parallel (as in Equation (5)) and the vectors connecting the centroids of adjacent planar segments should be perpendicular to the normal vectors (as in Equation (6)).

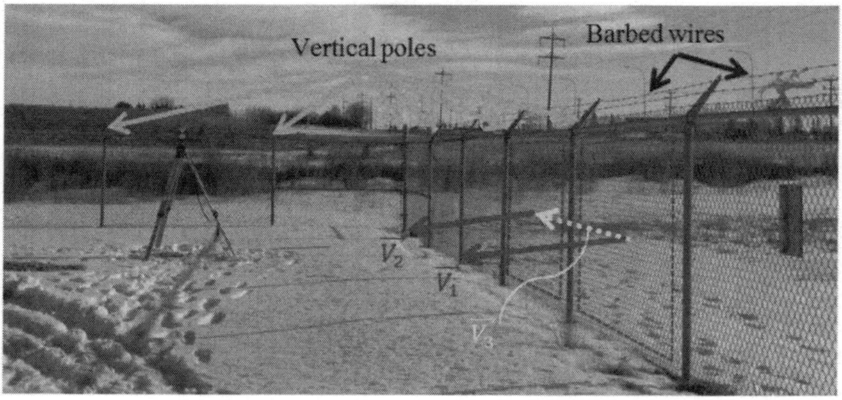

Figure 5. Fence recognition by detecting adjacent parallel planar segments, separated by vertical poles. The dashed polygons rectangles represent two planar segments.

$$\vec{V}_1 \| \vec{V}_2 \quad (5)$$

$$\vec{V}_1 \perp \vec{V}_3 \text{ and } \vec{V}_2 \perp \vec{V}_3 \quad (6)$$

where $\vec{V}1$ and $\vec{V}2$ denote the normal vectors of planar segments and $\vec{V}3$ is the vector that connects their centroids. Afterwards, region growing is employed to recognize the remaining parts of the fence (poorly-sampled panels and vertical poles) and also to agglomerate the recognized segments. The region growing algorithm then randomly chooses a seed point from the identified fence points and clusters points within 2-m 3D Euclidean distance. This process is repeated until no more un-segmented points remain within the 2-m threshold. The region growing distance threshold was selected based on prior knowledge of the required safety clearance between the fence and other electrical elements (Figure 6). The result is that all points belonging to fence are clustered in one segment.

4.2.2. Second Algorithm for Fence Recognition

This algorithm recognizes the fence based on its planimetric extent. First, all points are projected onto the horizontal plane. Since the fence is vertically distributed, the projected points appear as a linear feature. Other substation objects, such as cables and horizontally oriented bus pipes, also appear as linear objects once projected in this way. However, since the fence encloses the substation, it has the largest planimetric areal extent among all objects on a substation site, as is evident in Figure 6. Next, 2D eigenvalue analysis is performed to identify points belonging to a linear neighborhood, i.e.,

$$n\lambda_1 = 1 \ and \ n\lambda_2 = 0 \quad (7)$$

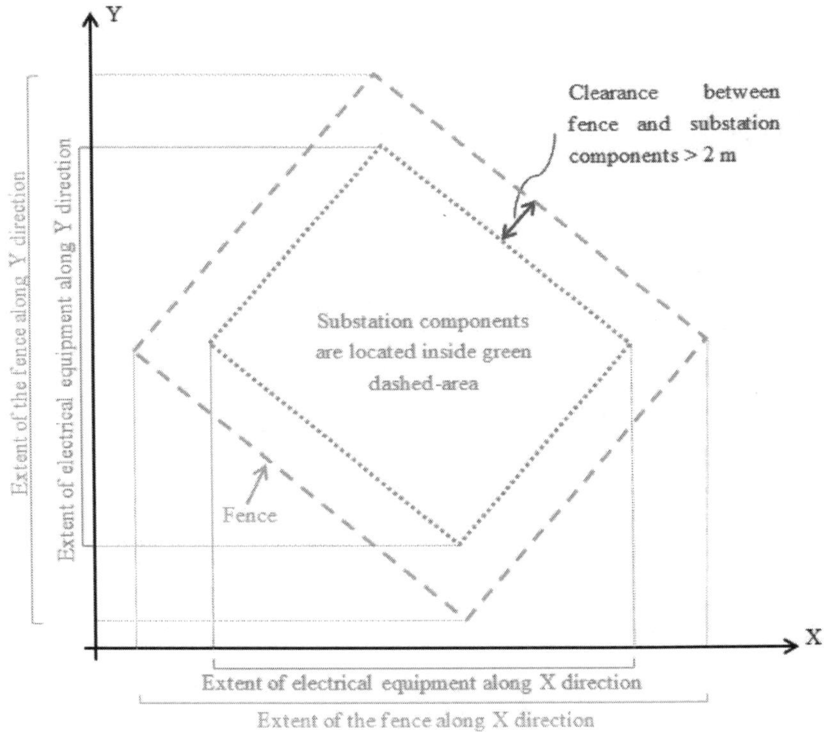

Figure 6. The clearance between the fence and other components in a substation site (plan view).

Region growing commences by randomly choosing one of points belonging to a linear neighborhood as a seed point. The algorithm then agglomerates all linear-neighborhood points within the 2-m 2D Euclidean distance of the seed point until no linear neighborhood remain. The result is a set of linear segments and the longest linear segment defines the fence.

4.3. Cable Recognition

Cables and some incompletely-scanned poles are two types of linear objects that appear in electrical substation LiDAR data. Cables appear as curvilinear objects while partially-scanned poles appear as straight linear objects. Thus, as the flowchart indicates in Figure 7, linear objects are first identified and then curvilinear objects are differentiated from straight linear objects.

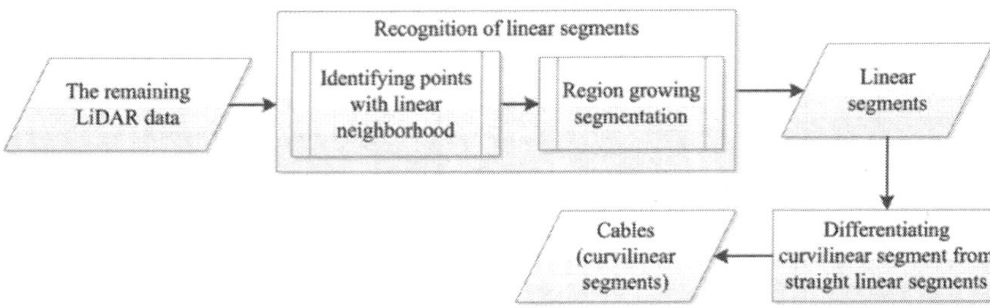

Figure 7. Flowchart of cable recognition.

First, eigenvalue analysis is performed to detect points belonging to linear neighborhoods. The following condition is used to identify points belonging to a linear neighborhood:

$$n\lambda_1 = 1 \; and \; n\lambda_2 = n\lambda_3 = 0 \qquad (8)$$

Afterwards, a region-growing algorithm similar to the algorithm used in the second fence recognition algorithm is employed to segment points belonging to a linear neighborhood. The result is a set of linear segments that include cables and non-cable straight linear segments. Cables are distinguished on the basis of their curvature. A cable and a non-cable straight linear segment are shown in Figure 8 in which red points indicate sample query points, red circles represent their 0.15 m 3D spherical neighborhoods and red arrows point to their principal direction vectors estimated by $\vec{v}$ 1 (eigenvector corresponding to the largest normalized eigenvalue). As is evident in this figure, the principal direction vectors along the cable vary from point to point (representing the cable's curvature) but they do not vary along the pole (indicating the pole's straightness).

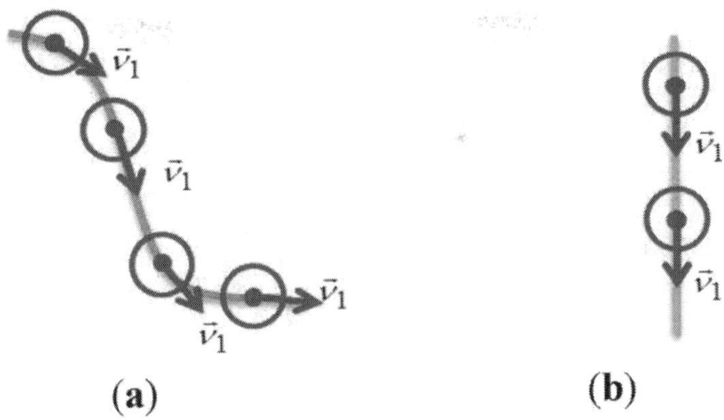

Figure 8. Principal orientation vectors along a cable (**a**) and along a pole (**b**). The arrows point to the principal direction within local neighborhoods.

The standard deviations of three components of $\vec{v}1$, i.e., (v_x, v_v, v_z), for all points of each linear segment are used to distinguish between cables and poles. Poles have nominally zero-valued standard deviations in all three dimensions (Equation (9)) since there is no variation in the direction vectors on a pole. Cables have at least one non-zero standard deviation along one of the three cardinal directions, as in Equation (10).

$$\sigma_{v_x} = \sigma_{v_y} = \sigma_{v_z} = 0 \tag{9}$$

$$\sigma_{v_x} \neq 0 \; and/or \; \sigma_{v_y} \neq 0 \; and/or \; \sigma_{v_z} \neq 0 \tag{10}$$

where σvx, σvy and σvz are standard deviation of $\vec{v}1$ three components in the three cardinal directions.

4.4. Recognition of Circuit Breakers and Bushings

The remaining planar segments identified in the fence recognition step are further analyzed to detect the circuit breakers and the bushings. The flowchart of circuit breaker and bushing recognition is given in Figure 9.

The topological structure common to both types of circuit breakers is one plane bounded by two adjacent perpendicular planes, as shown in Figure 10. Two following criteria are employed to test if such a structure exists:

$$\vec{V}_1 \perp \vec{V}_2 \text{ and } \vec{V}_1 \perp \vec{V}_3 \tag{11}$$

$$\vec{V}_4 \parallel \vec{V}_2 \text{ and } \vec{V}_4 \parallel \vec{V}_3 \tag{12}$$

If both these criteria are satisfied, then the three planar segments are grouped into one segment and identified as a circuit breaker. Afterwards, region growing is used to recognize the un-recognized parts of the circuit breakers: the non-planar parts; the fourth plane of a regular circuit breaker; and the cylindrical part of an oil circuit breaker. The recognized points on circuit breakers are considered as seed points. Seed points are randomly chosen and points within 0.15 m 3D Euclidean distance are clustered. The growing continues until no more un-segmented point is left in close neighborhood of seed points. The region growing does not include any points belonging to other objects since the only nearby objects are bushings above them, which are topped by cables. Recall that the ground is removed in the first step and cables are recognized in the second step. As a result, all points on each circuit breaker and bushings on top are aggregated in one segment.

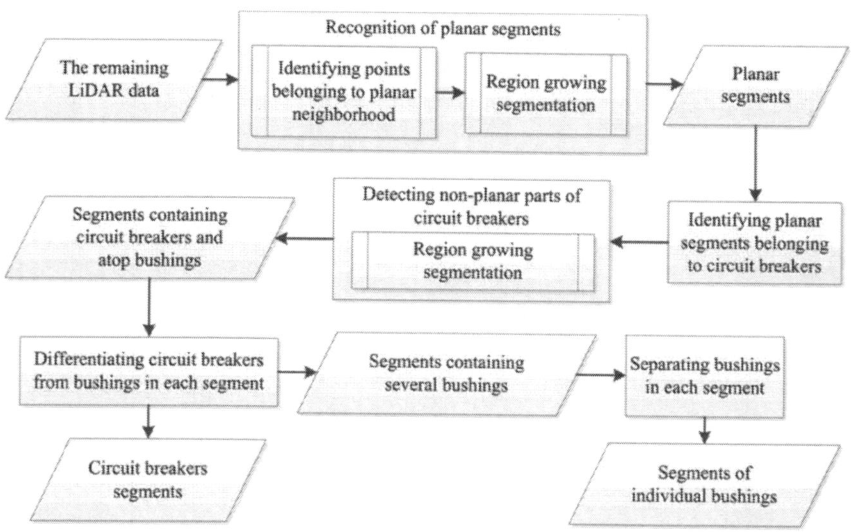

Figure 9. Flowchart of circuit breaker and bushing recognition.

Bushings, which sit atop a circuit breaker, are then separated on the basis of height. A circuit breaker's body, legs and bushings are identified by inspecting the height histogram. Each segment is divided into 0.1 m wide slices along the vertical direction (horizontal cross section). The bin size needs to be larger than the height accuracy (0.005 m in this case) so that the local point density is correctly reflected in the histogram. Figure 11a presents an example of a segment containing a circuit breaker and six bushings and Figure 11b shows the corresponding histogram. A few points on the topmost part of the circuit breakers planar parts may be incorrectly assigned to the bushing segments.

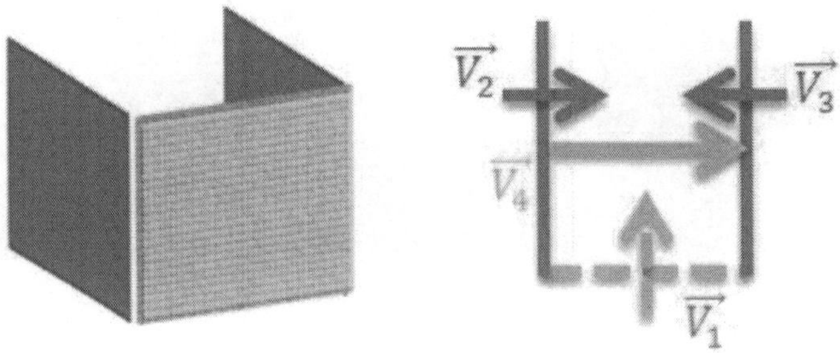

(a) (b)

Figure 10. The structure common in both two types of circuit breakers consisting of a plane with two adjacent planes in oblique view (**a**) and plan view (**b**). The normal vectors of the plane under study and the two adjacent planes are denoted by $\vec{V}1$, $\vec{V}2$ and $\vec{V}3$, respectively. The vector connecting the centroids of two adjacent planes is denoted by $\vec{V}4$.

The bushings are separated from one another and from the circuit breaker edges. Bushings are attached to the circuit breaker at the bottom and are separated from each other by a certain clearance (greater than 0.1 m) as required by safety restrictions. Their point clouds are predominantly vertically distributed while points on circuit breakers' top edges are mainly horizontally distributed. Therefore, the extents of each point's neighborhood along three cardinal directions are calculated. The largest extent represents the primary distribution direction and the horizontally distributed neighborhoods (belonging to circuit breakers' top edges) are discarded. Next, points belonging to vertically distributed neighborhoods that are within 0.05 m 3D Euclidean distance of one another are clustered so that each bushing belongs to a separate segment (Figure 12). The distance threshold (0.05 m) is selected considering the clearance among bushings (greater than 0.1 m).

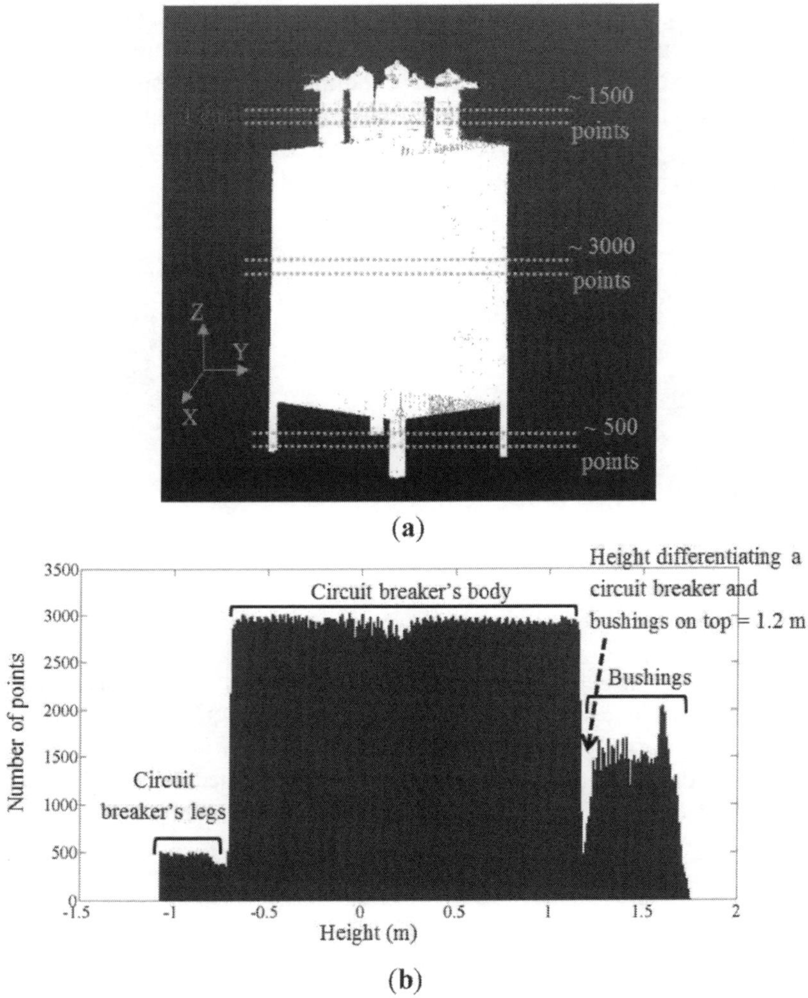

Figure 11. A segment containing a circuit breaker topped with six bushings is divided to slices in vertical direction (**a**). Histogram of slices' points number (**b**).

4.5. Bus Pipe Recognition

Bus pipes in electrical substations appear as cylindrical-shaped poles whose length is much larger than their radius. Their length easily reaches to several meters while their radius is approximately 0.04 m. First, eigenvalue analysis is employed to detect points belonging to a cylindrical neighborhood. Since bus pipes are mostly distributed along their main axis, the largest normalized eigenvalue ($n\lambda_1$) is much larger than other two normalized eigenvalues ($n\lambda_2$,

$n\lambda_3$) and its corresponding eigenvector ($\vec{v}1$) defines the cylinder axis (Figure 13). A point is considered to belong to a cylindrical neighborhood if:

$$n\lambda_1 > 0.5 \text{ and } n\lambda_2 = n\lambda_3 \tag{13}$$

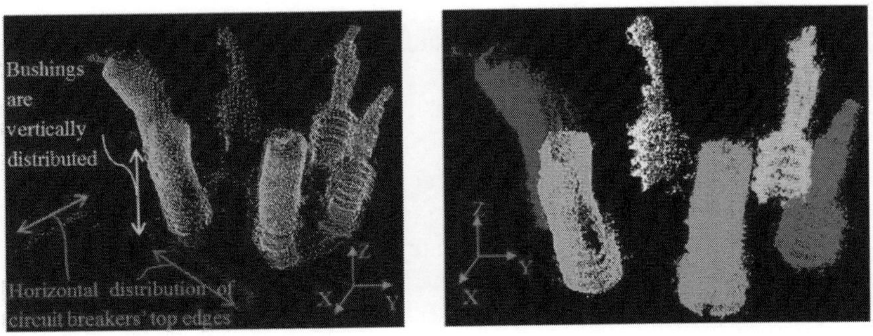

(a) (b)

Figure 12. Bushings (yellow) are vertically distributed while circuit breakers edges (blue and green) are distributed in horizontal directions (**a**). Each individually recognized bushing is indicated by a different color (**b**).

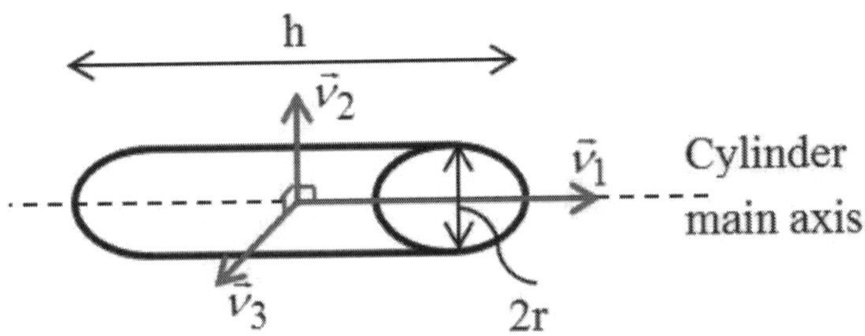

Figure 13. Eigenvectors within cylindrical neighborhood of bus pipes, considering h >> r.

Region growing segments the points identified as belonging to bus pipes by randomly selecting seed points from the set of points having a cylindrical-shaped neighborhood. The following three criteria are used for the aggregation:

$$\vec{v}1_{seed} \parallel \vec{v}1_{candidate} \tag{14}$$

$$\vec{v}1_{\text{seed}} \parallel \vec{V}_{\text{candidate}} \tag{15}$$

$$|h_{\text{seed}} - h_{\text{candidate}}| \leq \text{bus pipes radius} \tag{16}$$

where:

$$\vec{V}_{\text{candidate}} = (X_{\text{candidate}} - X_{\text{seed}}, Y_{\text{candidate}} - Y_{\text{seed}}, Z_{\text{candidate}} - Z_{\text{seed}}) \tag{17}$$

Equation (14) tests if the principal direction vectors ($\vec{v}1$) of the seed point and a candidate point are parallel. Equation (15) stipulates that the vector connecting the seed point to a candidate point is parallel to $\vec{v}1$ of the seed point. Equation (16) checks whether the height of a candidate point is almost the same as the height of the seed point.

If a candidate point satisfies all three above conditions, it is clustered and this process is pursued until no more un-segmented points belonging to a cylindrical neighborhood is left. As a result, each cylindrical pole belongs to one segment. The thresholds used in Equations (13) and (16) are selected based on standard bus pipes dimensions. Figure 14 illustrates bus pipe segmentation process in which only CP_2 is merged to the growing segment.

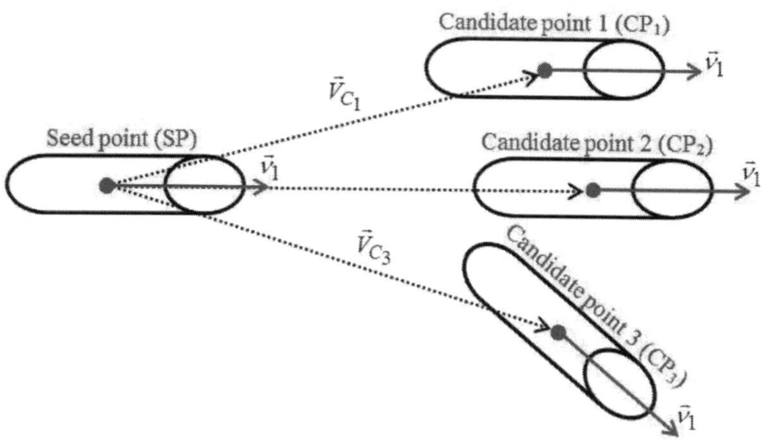

Figure 14. Bus pipe segmentation process in which all points (incorporating the seed point (SP) and three candidate points (CP_i)) are assumed to belong to cylindrical neighborhoods. CP_1 is not merged to the growing segment since although it satisfies the first condition (Equation (14)), it does not meet the condition in Equation (15). CP_2 is segmented as it meets all criteria. CP_3 is also not merged because neither the first condition (Equation (14)) nor the second condition (Equation (15)) is satisfied.

5. DATASETS

5.1. First Dataset

The first dataset was acquired by scanning an electrical substation located in Airdrie, north of Calgary, Canada using a Leica HDS 6100 terrestrial laser scanner, shown in Figure 15a. Scanning was carried out from seven locations to cover a portion of interest of the substation where the majority of electrical equipment was located. Only this portion, which covers a 75 m × 30 m area, was employed for further processing. Registration was performed using flat, signalized targets (Figure 15c). This resulted in a point cloud with more than 58 million points (Figure 16), which is color-coded according to intensity (a quantized measure of the reflected laser beam power). The black "holes" in Figure 16 indicate the instrument locations.

(a) (b) (c)

Figure 15. Leica HDS 6100 (**a**), Faro Focus^{3D} (**b**) and a flat signalized target (**c**).

The positional accuracy of this scanner is quite high (0.005 m), which suggests the suitability of the acquired data for dimensioning purposes. The primary error source in this dataset was multipath i.e., when the emitted laser beam returns to the laser scanner by two or more paths rather than a single straight path. Multipath is primarily caused by highly reflective surfaces. The multipath level is quite minimal i.e., less than two hundred points out of fifty eight million points. The multipath-introduced noise appears as points lying much higher or much lower than the ground, as portrayed in Figure 17a. The subset of points below the ground in a circular pattern around the instrument locations (Figure 17b suggests that the instrument's tripod legs might have caused the multipath since they have highly reflective surfaces. The source of multipath-introduced noise indicated by green in Figure 17a is unknown since it does not have a systematic pattern and also the majority of electrical components (like bus pipes) are usually galvanized and thus, do not have reflective surfaces.

Figure 16. First dataset, scanned by a Leica HDS 6100 terrestrial laser scanner.

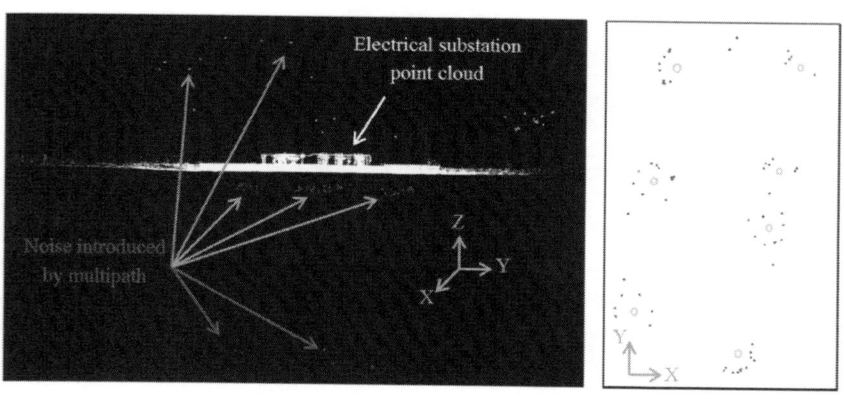

Figure 17. Multipath-introduced noise is indicated by red, blue and green colors based on their height (**a**). Noise follows a circular pattern around the instrument tripod locations (plan view). Small yellow circles represent the instrument's locations during scan time (**b**).

5.2. Second and Third Datasets

The second data set was captured by scanning the same electrical substation in Airdrie with a Faro Focus^{3D} terrestrial laser instrument (Figure 15b) from the same locations as the Leica scanner. The Faro Focus^{3D} scanner has a range accuracy of 0.002 m. Like the first dataset, only the portion of the substation was of interest and scanned and registration was performed using signalized targets. The resulting point cloud contains more than four million points, which is shown in Figure 18 and colored according to RGB data captured by the scanner's integrated camera. Although the first and second datasets cover the same area, the second dataset was intentionally captured with much lower point sampling so that the impact of point sampling on the proposed methodology can be studied. That is why the first and second datasets have the same configuration but different point sampling.

The third dataset (Figure 19) was a pre-registered Faro Focus^{3D} point cloud of eight scans and 356 million points that was obtained from [32]. The third dataset volume is the largest among all datasets since it covers an entire electrical substation. RGB and intensity information are used in the following figures only for visualization purposes and only geometrical information (points' 3D co-ordinates) were used for data processing. The two Faro datasets do not suffer from multipath-introduced noise like the first dataset. Unlike the first dataset that was collected by the authors, the second and third datasets were acquired by other individuals who likely eliminated the noise in a pre-processing step.

Figure 18. Second dataset, scanned by a Faro Focus terrestrial laser scanner.

Table 2 summarizes the specifications of three datasets in terms of the employed laser scanner, the number of points, dimensions of the covered area, and the nominal point spacing.

Figure 19. Third dataset, scanned by a Faro Focus terrestrial laser scanner.

Table 2. Specifications of three datasets used in this research.

Specifications	1st Dataset	2nd Dataset	3rd Dataset
Scanner	Leica HDS 6100	Faro Focus^{3D}	Faro Focus^{3D}
Number of points	58,143,195	4,286,739	356,396,779
Area (m × m)	75 × 30	75 × 30	71 × 50
Nominal point spacing at 10 m (mm × mm)	3.1 × 3.1	10 × 10	10 × 10

Table 3 presents the number and the average point sampling on various types of objects in three datasets. The cables and bus pipes samplings are stated as the number of points per meter considering their much larger length compared to their thickness. The point sampling of circuit breakers and fences are presented as points per square meters considering their planar shape. The bushings' sampling is expressed as points per object considering their complex shape and small dimensions.

The point sampling of all objects in the third dataset is lower than those in the first and second dataset, even though it occupies the largest volume (the only exception is the fence point sampling which is further discussed in the next section). That is since a large portion of the third dataset is composed of points on objects that are not of interest in this study such as the control house and the distribution bus (the steel structure in the right side of Figure 19). These objects are not of interest in this work since they have no impact on occurrence or prevention of wildlife-related power outages, which is the scope of this work. For instance, the distribution bus is utilized only to hold electrical equipment in place and the control house is utilized to keep spare electrical equipment. Furthermore, in the third dataset, cables and circuit breakers lie in a very close proximity of the distribution bus. The dense structure of the distribution bus introduces occlusions and results in point sampling reduction of cables and circuit breakers. Occlusion is a shadow effect that is inevitable when the configuration of objects on site is dense. As a result, the third dataset has the largest volume among three datasets while the point sampling of its key elements are lower than that of the other two datasets.

Table 3. The number and average point sampling of each type of object in three datasets.

Objects	1st Dataset		2nd Dataset		3rd Dataset	
	Number	Sampling	Number	Sampling	Number	Sampling
Fence (points/m^2)	1	33,351	1	326	1	887
Cables (points/m)	69	2982	69	845	117	664
Circuit breakers (points/m^2)	6	87,940	6	14,806	7	8272
Bushings (points/object)	36	18,428	36	3472	39	1443
Bus pipes (points/m)	69	9979	69	2143	27	2196

6. RESULTS AND DISCUSSION

Figure 20 shows the recognized objects in the three datasets. The figure is colorized according to objects' type. Figure 21 present a finer-level view of the recognized objects in three datasets. Each color in these figures represents a separate segment identified as a circuit breaker, a bushing, a cable or a bus pipe. Moreover, as these figures show, points on the ground in all three datasets have been successfully eliminated.

To assess the results, objects of interest were manually delineated from the original point clouds and were used as ground truth data (presented in Table 3). The results assessment is presented in Table 4 in terms of the recognition accuracy and precision at the point cloud level and in Table 5 in terms of the number of the recognized objects at the object level. Accuracy and precision are calculated as follows.

$$\text{accuracy} = \frac{tp + tn}{tp + tn + fp + fn} \tag{18}$$

$$\text{precision} = \frac{tp}{tp + fp} \tag{19}$$

Where tp, tn, fp and fn denote the number of true positive, true negative, false positive and false negative occurrences, respectively.

(a)

(b)

(c)

Figure 20. Recognized object types in the first dataset (**a**), second dataset (**b**) and third dataset (**c**). Each color represents a different type of object.

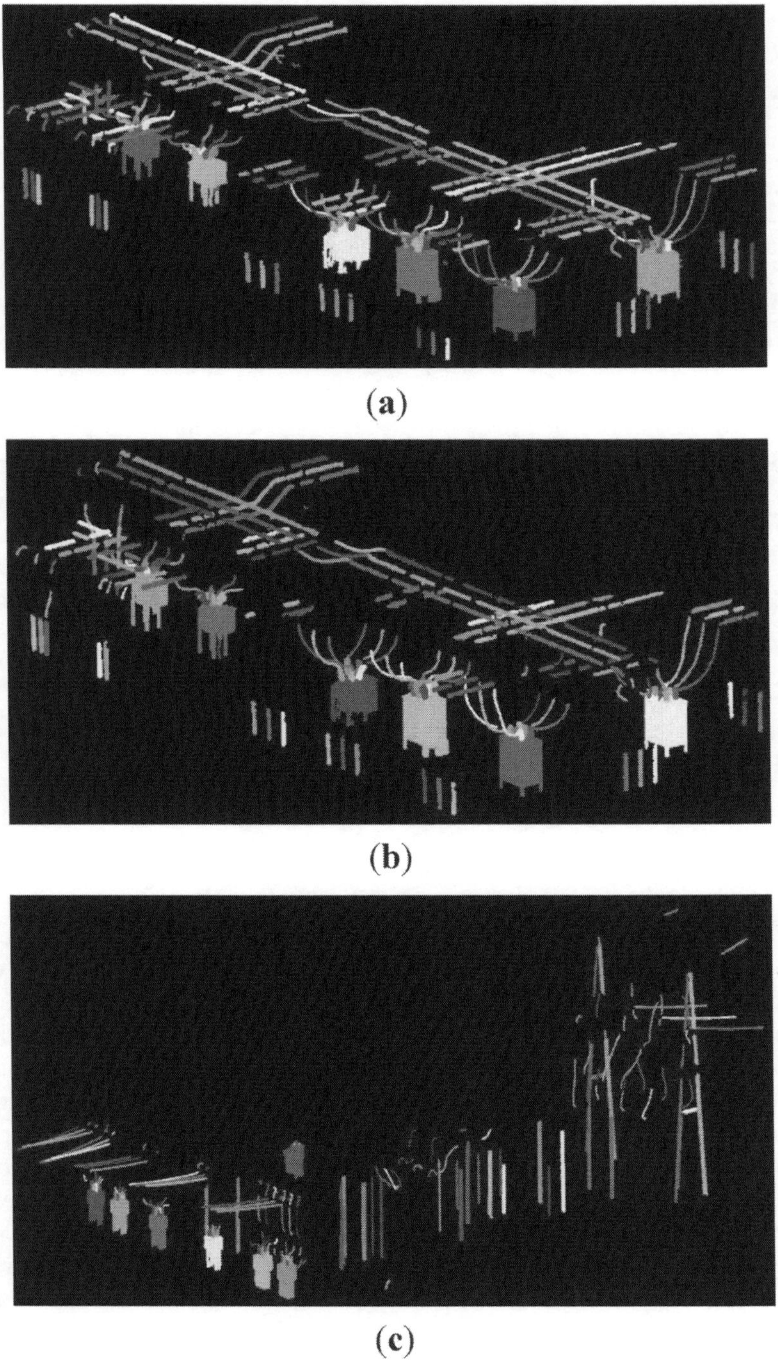

Figure 21. A finer-level of representation of recognized objects in the first dataset (**a**), second dataset (**b**) and third dataset (**c**). Each color represents a separate object that can be of any type.

Table 4. Accuracy and precision of object recognition at point cloud level (in percentage).

Objects	1st Dataset		2nd Dataset		3rd Dataset	
	Accuracy	Precision	Accuracy	Precision	Accuracy	Precision
Fence	99.7	100	98.1	100	96.7	100
Cables	98.5	97.6	92.8	90.66	93.2	93.78
Circuit breakers	100	100	100	100	100	100
Bushings	100	100	100	100	100	100
Bus pipes	98.2	100	92.8	100	97.1	100
Mean	99.28	99.52	96.74	98.13	97.4	98.76

Table 5. Recognition results at object level.

Objects	1st Dataset		2nd Dataset		3rd Dataset	
	Recognized Objects	False Positives	Recognized Objects	False Positives	Recognized Objects	False Positives
Fence	1 (1)	0	1 (1)	0	1 (1)	0
Cables	68 (69)	1	64 (69)	3	109 (117)	2
Circuit breakers	6 (6)	0	6 (6)	0	7 (7)	0
Bushings	36 (36)	0	36 (36)	0	39 (39)	0
Bus pipes	67 (69)	0	64 (69)	0	27 (27)	0
All objects	178 (181)	1	171 (181)	3	183 (191)	2

The high mean recognition accuracy of the fence (98.2%) suggests that the majority of points belonging to fence are recognized. The highest fence recognition accuracy was achieved for the first dataset, which was anticipated due to the finer point sampling of the fence in this dataset. Although average point sampling on the fence was higher for the third dataset than for the second, the associated fence recognition accuracy was slightly lower (less than 2%). This is due to non-uniform point sampling of the fence in the third dataset: some fence panels were very well-sampled while others were very poorly-sampled. Figure 20c shows an instance of such a poorly-sampled area with large gaps and, as a result, many fence panels were not recognized. Moreover, Figure 22 depicts some poorly-sampled fence panels suggesting a linear structure while its true structure is planar. Figure 22 also demonstrates fence recognition failure in a part of data with gaps larger than the 2-m region growing distance threshold. The high recognition accuracy of the sampled portions (especially in the first and second datasets) demonstrates that if the fences were well-sampled, they can be successfully recognized. Furthermore, there are no false positives in the results of fence recognition of all three datasets implying 100% recognition precision. This demonstrates the suitability of using fence's unique shape (i.e., multiple adjacent planar surfaces) for its recognition.

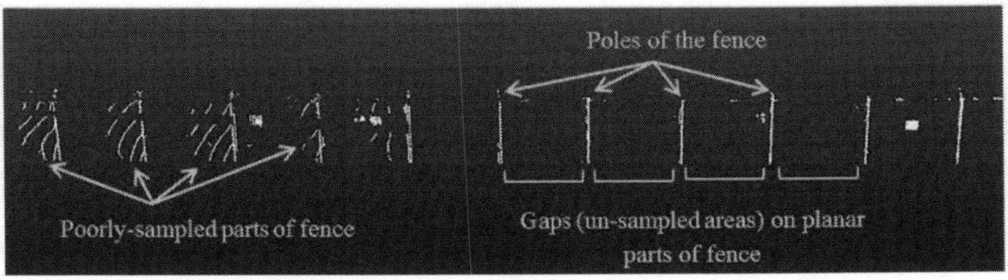

Figure 22. Poorly-sampled parts and gaps (un-sampled parts) of the fence in the third dataset.

The cable recognition was also very successful as 98.5% of cables in the first dataset were recognized and there was only one false positive. The cables of the second and third datasets were recognized with greater than 92% accuracy. However, there were three false positives in the second dataset and two false positives in the third. The false positives were caused by under-sampled poles that appear as curvilinear objects in the point clouds and were incorrectly recognized as cables. The highest cable recognition accuracy and precision was achieved for the first dataset, which is expected since the associated cable point sampling was the highest among three datasets. Although cable recognition accuracy of the third dataset is quite high (greater than 93%), it is the lowest among three datasets since the majority of cables are situated among the distribution bus components. The distribution bus (as is evident in Figure 19) has a very dense structure and, as a result, many cables were highly occluded.

The recognition results of the circuit breaker and bushings indicate that all were recognized in all three datasets. There are also no false positives in the results of three datasets, so the recognition precision was 100%. The isolated position of the circuit breakers and bushings has a great impact on their perfect recognition results since they were well-sampled by the laser scanner and occlusions by other objects were quite minimal. This is evident in Table 3 that indicates circuit breakers have the highest point sampling among all key components in each dataset. Furthermore, exploiting the bushings' distinctive topology (i.e., locating between circuit breakers and cables) rather than using their shape for their recognition lead to perfect results, even though they have the most complicated physical shape among all key components of electrical substations.

As indicated in Table 5, 67 and 64 out of 69 bus pipes in the first and second datasets were recognized. Again, the highest bus pipe recognition accuracy (98.2%) at the point cloud level was realized for the first dataset due to the highest sampling. In the third dataset, all 27 bus pipes were recognized (100% accuracy at the object level) and 97.1% accuracy at the point cloud level was obtained. Although the bus pipe point-sampling of the second and third datasets was almost the same, results are more successful in the third dataset. This stems from denser configuration of bus pipes in the second dataset, which is evident in Figure 21b,c. Furthermore, there were no false positives in the bus pipe recognition results for all three datasets.

As the results suggest, under-sampling, caused by sparse point coverage and/or occlusions, is the primary cause of the recognition failure and false positives. When a large portion of an object is under-sampled, its form and structure cannot be accurately recognized according to the described rules. However, it is crucial to note that the determining factor for the recognition success is the point sampling of the objects of interest and that considering only the total volume of the datasets in evaluating the results can be misleading. This is evident in the results of the third dataset that has the largest volume among all three datasets while the point sampling of its key components is the lowest. Occlusions and sparse sampling can be ameliorated by acquiring more scans from various locations. This action, however, requires additional time for fieldwork and results in a point cloud with larger volume, which increases computational time and cost and decreases data processing efficiency. Therefore, a balance is required between the number of scan stations and size of the dataset to achieve acceptable recognition results with a reasonable processing time and cost. This balance can be achieved by considering the number, size and configuration of components at an electrical substation site. Furthermore, the local neighborhoods are characterized using PCA, which can be sensitive to noise. As explained in Section 4, the noise level in the datasets used in this work was quite minimal. However, PCA might not produce reliable information regarding the local neighborhood structure for a dataset with a higher noise level. In such a case, employing RANSAC is recommended.

7. CONCLUSIONS

In this study, the automated recognition of key components of electrical substations from 3D LiDAR datasets is proposed in support of model construction and dimensional extraction. To this end, a novel methodology (composed of six different algorithms) was proposed to recognize the objects of interest. Three datasets with differing volumes, point spacing and configuration captured with two different terrestrial laser scanners were employed. The proposed methodology achieved an average accuracy of 97.8% and an average precision of 98.8% at the point cloud level. Furthermore, 178 and 171 out 181 objects in the first and second datasets and 183 out of 191 elements in the third dataset were successfully identified. The successful object recognition from electrical substation point clouds with various volumes, configuration and point sampling indicates the strength and robustness of the proposed methodology. The achieved results suggest that automated recognition of equipment in electrical substations by using 3D LiDAR datasets is feasible and promising. The results also provide very valuable information regarding the current state of the substations without manual inspection of every single piece of equipment by staff. The results of this work can be used to obtain precise dimensions of electrical equipment to manufacture custom-built, non-conductive covers that reduce wildlife-related power outages.

8. OUTLOOK

The performance of the proposed methodology can be studied for data with high noise level. In such case, authors recommend replacing PCA by RANSAC to study the point neighborhood structure since RANSAC is less sensitive to noise, compared to PCA. The results of this study can be used to create as-built models of electrical substations by fitting 3D CAD models to the recognized objects. As-built 3D representations provide precise dimensions of electrical equipment that are important for the manufacturing of custom-built insulating covers.

ACKNOWLEDGMENTS

Tecterra, the Canada Foundation for Innovation (CFI) and the Antje Graupe Pryor Foundation are acknowledged for their financial support. Cantega Technologies is also highly acknowledged for their support and arranging access to an electrical substation in Airdrie Canada for data collection.

AUTHOR CONTRIBUTIONS

The research was conducted by Mostafa Arastounia under the direct supervision of Derek D. Lichti. The tasks for the research include the design and implementation of the proposed methods, data collection, and result analyses. Both authors have drafted, edited and reviewed the manuscript.

CONFLICTS OF INTEREST

The authors claim no conflict of interest.

REFERENCES

1. USA Department of Energy. Available online: http://energy.gov/sites/prod/files/oeprod/DocumentsandMedia/DOE_SG_Book_Single_Pages(1).pdf (accessed on 22 November 2014).
2. James, J.B. Factors Affecting Wildlife-Related Power Outages in Electrical Substation. Master's Thesis, Oklahoma State University, Stillwater, OK, USA, 1998.
3. TD World. Available online: http://tdworld.com/substations/reliability-takes-wing (accessed on 14 September 2015).
4. Burnham, J.; Carlton, R.; Cherney, E.A.; Couret, G.; Eldridge, K.T.; Farzaneh, M.; Frazier, S.D.; Gorur, R.S.; Harness, R.; Shaffner, D.; et al. Preventive measures to reduce bird-related power Outages-part I: Electrocution and collision. IEEE Trans. Power Deliv. **2004**, 19, 1843–1847.
5. Sundararajan, R.; Burnham, J.; Carlton, R.; Cherney, E.A.; Couret, G.; Eldridge, K.T.; Farzaneh, M.; Frazier, S.D.; Harness, R.; Shaffner, D.; et al. Preventive measures to reduce bird related power outages-part II: Streamers and contamination. IEEE Trans. Power Deliv. **2004**, 19, 1848–1853.
6. New Mexico Avian Protection. Available online: http://nmavianprotection.org/wp-content/uploads/2012/02/Power-Lines-

Substations-Problems-Mitigation-Strategies_Harness.pdf (accessed on 18 September 2014).

7. Sternberg, H.; Kersten, T.P. Comparison of terrestrial laser scanning systems in industrial as-built documentation applications. Opt. 3-D Meas. Tech. **2007**, 8, 389–397.

8. Xu, Z.; Yong, F.; Huang, Y.; Wang, Z.; Liu, Y. LiDAR applications in the electrical power. Int. Arch. Photogramm. Remote Sens. Spat. Inf. Sci. **2008**, 36, 137–140.

9. Gonzalez-Aguilera, D.; Del Pozo, S.; Lopez, G.; Rodriguez-Gonzalvez, P. From point cloud to CAD models: Laser and optics geotechnology for the design of electrical substations. Opt. Laser Tech. **2012**, 44, 1384–1392.

10. Armeshi, H.; Habib, A. Automated extraction of ceramic insulators from imagery of electrical substations. In Proceedings of the International Symposium on Mobile Mapping Technology, Tainan, Taiwan, 1–3 May 2013.

11. Arastounia, M.; Lichti, D. Automatic extraction of insulators from 3D LiDAR data of an electrical substation. ISPRS Ann. Photogramm. Remote Sens. Spat. Inf. Sci. **2013**, 2, 19–24.

12. Vosselman, G.; Gorte, B.G.H.; Sithole, G.; Rabbani, T. Recognizing structure in laser scanner point clouds. Int. Arch. Photogramm. Remote Sens. Spat. Inf. Sci. **2004**, 36, 33–38.

13. Hough, P.V.C. Method and Means for Recognizing Complex Patterns. U.S. Patent 3069654, 18 December 1962.

14. Belton, D.; Lichti, D. Classification and segmentation of terrestrial laser scanner point clouds using local variance Information. Iaprs **2006**, 5, 44–49.

15. Rabbani, T.; Van Den Heuvel, F.A.; Vosselman, G. Segmentation of point clouds using smoothness constraint. Int. Arch. Photogramm. Remote Sens. Spat. Inf. Sci. **2006**, 36, 248–253.

16. Pu, S.; Vosselman, G. Automatic extraction of building features from terrestrial laser scanning. Int. Arch. Photogramm. Remote Sens. Spat. Inf. Sci. **2006**, 36, 25–27.

17. Pu, S.; Vosselman, G. Extracting windows from terrestrial laser scanning. Int. Arch. Photogramm. Remote Sens. Spat. Inf. Sci. **2007**, 36, 320–325.

18. Pu, S.; Vosselman, G. Knowledge based reconstruction of building models from terrestrial laser scanning data. ISPRS J. Photogramm. Remote Sens. **2009**, 64, 575–584.

19. Lehtomäki, M.; Jaakkola, A.; Hyyppä, J.; Kukko, A.; Kaartinen, H. Detection of vertical pole-like objects in a road environment using vehicle-based laser scanning data. Remote Sens. **2010**, 2, 641–664.

20. Martínez, J.; Soria-Medina, A.; Arias, P.; Buffara-Antunes, A.F. Automatic processing of terrestrial laser scanning data of building façades. Autom. Constr. **2012**, 22, 298–305.

21. Fischler, A.; Bolles, C. Random sample consensus: A paradigm for model fitting with applications to image analysis and automated cartography. Commun. ACM **1981**, 24, 381–395.

22. Arastounia, M.; Lichti, D. Segmentation of planar surfaces in LiDAR point clouds of an electrical substation by exploring the structure of points' neighborhood. ISPRS Int. Arch. Photogramm. Remote Sens. Spat. Inf. Sci. **2014**, 40, 55–62.

23. Zhu, L.; Lehtomaki, M.; Hyyppa, J.; Puttonen, E.; Krooks, A.; Hyyppa, H. Automated 3D scene reconstruction from open geospatial data sources: Airborne laser scanning and a 2D topographic database. Remote Sens. **2015**, 7, 6710–6740.

24. Melzer, T.; Briese, C. Extraction and modeling of power lines from ALS point clouds. In Proceedings of the 28th Workshop of the Austrian Association for Pattern Recognition, Hagenberg, Austria, 17–18 June 2004; pp. 47–54.

25. Jwa, Y.; Sohn, G.; Kim, H.B. Automatic 3D power line reconstruction using airborne LiDAR data. Int. Arch. Photogramm. Remote Sens. **2009**, 38, 105–110.

26. Jwa, Y.; Sohn, G. A multi-level span analysis for improving 3D power-line reconstruction performance using airborne laser scanning data. Int. Arch. Photogramm. Remote Sens. Spat. Inf. Sci. **2010**, 38, 97–102.

27. Sohn, G.; Jwa, Y.; Kim, H.B. Automatic power line scene classification and reconstruction using airborne Lidar data. ISPRS Ann. Photogramm. Remote Sens. Spat. Inf. Sci. **2012**, 13, 167–172.

28. Zhang, J.; Lin, X.; Ning, X. SVM-based classification of segmented airborne LiDAR point clouds in urban areas. Remote Sens. **2013**, 5, 3749–3775.

29. Zhu, L.; Lehtomaki, M.; Hyyppa, J. Fully-automated power line extraction from airborne laser scanning point clouds in forest areas. Remote Sens. **2014**, 6, 11267–11282.

30. Zhu, L.; Hyyppa, J. The use of airborne and mobile laser scanning for modeling railway environments in 3D. Remote Sens. **2014**, 6, 3075–3100.

31. USA Department of Labor. Available online: https://www.osha.gov/SLTC/etools/electric_power/illustrated_glossary/substation.html#Equipment (accessed on 14 October 2014).

32. USA LiDAR. Available online: http://www.lidarusa.com/index.php (accessed on 14 September 2014).

CHAPTER 5

Detection of Gold Nanoparticles Aggregation Growth Induced by Nucleic Acid through Laser Scanning Confocal Microscopy

Ramla Gary [1,*], Giovani Carbone [1], Gia Petriashvili [1,2], Maria Penelope De Santo [1,3] and Riccardo Barberi [1,3]

[1] Physics Department, University of Calabria, Rende 87036, Italy

[2] Institute of Cybernetics of the Georgian Technical University, Euli str. 5, 0175 Tbilisi, Georgia

[3] CNR-Nanotec UOS di Cosenza, c/o University of Calabria, Rende 87036, Italy

ABSTRACT

The gold nanoparticle (GNP) aggregation growth induced by deoxyribonucleic acid (DNA) is studied by laser scanning confocal and environmental scanning electron microscopies. As in the investigated case the direct light scattering analysis is not suitable, we observe the behavior of the fluorescence produced by a dye and we detect the aggregation by the shift and the broadening of the fluorescence peak. Results of laser scanning confocal microscopy images and the fluorescence emission spectra from lambda scan mode suggest, in fact, that the intruding of the hydrophobic moiety of the probe within the cationic surfactants bilayer film coating GNPs results in a Förster resonance energy transfer. The environmental scanning electron microscopy images show that DNA molecules act as template to assemble GNPs into three-dimensional structures which are reminiscent of the DNA helix. This study is useful to design better nanobiotechnological devices using GNPs and DNA.

KEYWORDS

gold nanoparticle aggregation; deoxyribonucleic acid; laser scanning confocal microscopy; surface plasmon resonance; Förster resonance energy transfer; hydrophobicity

1. INTRODUCTION

Individual particles and assemblies of gold nanoparticles (GNPs) show different optical and electrical properties. For instance, the color change due to aggregation of GNPs is an indicator of the surface plasmon shift, allowing potential applications in sensors to recognize DNA or its components [1].

Life processes strongly depend on nucleic acids and proteins which, therefore, are used as probes for measurements of other components in biological systems [2]. The development of nanobiotechnology is helped by a deep understanding of the interaction of GNPs with nucleic acids, which are used as a basic tool for nanobiotechnological devices [3,4]. Examples of this approach are the organization of metal and semiconductor nanoclusters [5], several bioanalytical techniques [6], the biomolecular electronics [7] and nanomechanical systems. Therefore, the DNA efficiency for inducing GNPs self-assembly is a very interesting subject of research. J. Yang et al. [8] demonstrated how single-stranded DNA could be used to manage the placement of two different sized GNPs in a complex particle system. K. N. Ganesh et al. [9] have studied electrostatic assembly of cationic surfactants-capped GNPs on DNA duplex. H. Nakao et al. [10] have reported the highly-ordered assemblies of GNPs in well-aligned and long-range order on DNA molecules.

Color change, UV-VIS spectroscopy, and dynamic light scattering have been used as techniques for the detection of GNPs aggregation [11,12,13]. Confocal and two photon microscopies and single-cell plasmonically-enhanced Rayleigh-scattering imaging spectroscopy were used as imaging techniques to observe the GNPs aggregation after their accumulation within living cells through endocytosic processes [14,15,16]. These approaches take advantage of the characteristic optical properties of plasmonic GNPs, namely, their ability to strongly scatter light, and the coupling of their plasmonic fields when particles come into close proximity.

In this work, the laser scanning confocal microscopy (LSCM) is used to observe the double-stranded DNA induced aggregate growth of cationic surfactants-capped GNPs. The observation based on the light scattering enhancement from GNPs aggregates is not recommended in our experiments since there is a strong overlap between the light scattering from DNA/cationic surfactants complexes and the light scattering from GNPs aggregates (see in Appendix). Therefore, to skirt this overlap problem we detected the GNP aggregation through the Förster Resonance Energy Transfer (FRET) process. Nile Blue perchlorate (NB) is used as a probe because, in addition to the NB molecules ability to be intercalated inside the DNA helix [17,18], it seems that this probe can be also entrapped within the hydrophobic part of the surfactants bilayer films coating the aggregated GNPs. This entrapment places this probe in proximity to metallic surfaces leading to FRET. Due to the strong interaction between NB molecules and the large localized fields induced by the plasmonic coupling, a highly enhanced fluorescence is produced allowing the localization of GNPs aggregates.

2. PREPARATION OF SAMPLES AND METHODS

2.1. Gold Nanoparticles

Our experiments are performed by using commercially available GNPs, obtained from Sigma Aldrich as stabilized suspension in citrate buffer with a core size of 37–43 nm and a mean hydrodynamic diameter of 48–56 nm. GNPs are usually prepared by using citrate reduction of Au ions in an aqueous solution; therefore, GNPs surface is negatively charged. To minimize their aggregation, the versatile surface chemistry of citrate-capped GNPs allows them to be coated with a hydrophilic bilayer film on their surface via electrostatic interaction. By using a stabilizer, such as cetyltrimethylammonium bromide (CTAB), which presents a positive hydrophilic ammonium head group and a hydrophobic long hydrocarbon tail, the surface electric charge is converted from negative to positive [19,20].

The absorption spectrum of the solution containing the dispersed GNPs (Figure 1) was observed through a UV-VIS-NIR spectrophotometer AVASPE-2048 Avantes. Surface plasmon resonance (SPR) causes strong scattering as well as absorption of light. Therefore, the absorption spectrum of dispersed GNPs provides data about the SPR band of GNPs. Our observations show that SPR band of the GNPs used for this work ranges from 400 nm to 600 nm, with a peak at 525 nm (Figure 1).

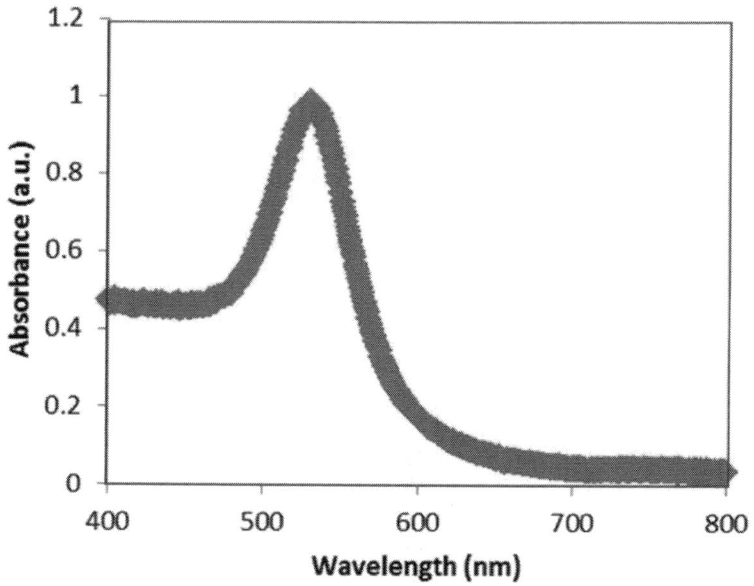

Figure 1. Absorption spectrum of GNPs stabilized suspension in citrate buffer.

2.2. Preparation of the DNA Solution

Nucleic acid solution was prepared by dissolving 10 mg of commercially available deoxyribonucleic acid (double-stranded (ds) molecule), DNA, fish sperm from salmon testes (Sigma-Aldrich) in 10 mL of doubly distilled water. 24 h or more were needed for the dissolution at 4 °C, accompanied by occasional gentle shaking.

2.3. LSCM Measurements

For the observation of the GNPs aggregation growth induced by DNA based on the NB fluorescence analysis, four solutions were prepared. The DNA/GNPs/NB solution is prepared at first. We added to 1.6 g of the GNPs stabilized suspension in citrate buffer, 0.002 g of the crystallized probe Nile Blue A perchlorate (NB; Figure 2) purchased from Sigma-Aldrich. This mixture is left in a shaker for 15 min at 50 °C in order to dissolve the probe. Then, 1.3 mL from the dissolved DNA was transported to the previous GNPs/NB mixture. The three other solutions were prepared for control experiments. The first one was prepared with the same procedure described above except that this mixture was free from DNA. To have the same concentration of GNPs in DNA/GNPs/NB solution described previously, 1.3 mL of double distilled water was added to GNPs/NB solution (solution No. 1 for control experiments).

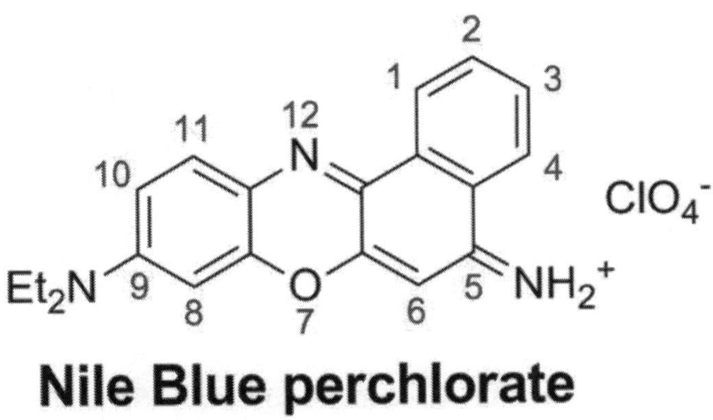

Figure 2. Molecular structure of the fluorescence probe NB.

To study the effect of GNPs on the fluorescence of NB bound to the cationic surfactants present as stabilizer in the GNPs suspension in citrate buffer, the solution No. 2 was prepared with the same procedure described above for the GNPs/NB solution (solution No. 1) except that GNPs were removed from the buffer solution by centrifugation at 3000 rpm for an hour. The final solution for control experiments (solution No. 3) was prepared by adding 0.002 g of NB to 1.3 mL from the dissolved DNA. Then, the resulting solutions were gently shaken at room temperature for 5 min, stored at 4 °C for 48 h, and 0.25 mL from each solution was deposited by drop-coating to the glass slide treated by deionized water. The four coated films on substrate were stored for 24 h at room temperature to let the water evaporate completely resulting to four precipitates.

(LSCM) was used to image the fluorescent emission on each of the four samples. All samples were imaged on a Leica TCS SP8 using a 25 × /0.95 numerical aperture water immersion objective and identical setting. The small droplet of the immersive water was placed on the opposite side of the coated film on the glass slide with the thickness of 1.13 mm. The 514 nm was used as pump beam and the fluorescent emission was sent simultaneously to 2 PMTs (PMT1 and PMT2) collecting respectively the light in the (550–600) nm and (600–750) nm ranges. The SPR of GNPs and the excitation spectrum of NB have respectively the (400–600) nm and (470–690) nm ranges [21]. Thus, 514 nm was chosen for the irradiation because this wavelength is within these two ranges.

Since the NB emission depends on the polarity of the environment which is red shifted with higher polarity, emitted light was detected in (550–600) nm and (600–750) nm ranges [17]. The use of two PMTs allowed the simultaneous acquisition of two images. In order to enhance the contrast, each image is colorized by an artificial color, according to the detection ranges of PMT1 and PMT2, where the green or the yellow areas are the expression of the fluorescence from NB bound to the hydrophobic part of cationic surfactants and the red or the grey domains are the expressions of NB bound to DNA.

2.4. Scanning Electron Microscopy

An environmental scanning electron microscope (ESEM) Quanta 400F installed at Nanotec-Cosenza (Italy) by FEI Europe B.V (Achtseweg Noord, 5, 5651 GG Eindhoven, The Netherlands) and FEI Italy (Srl Viale Bianca Maria, 23, 20122 Milano (Mi)) Companies (Milano, Italy) has been used to image the samples prepared with GNPs at pressure of 1 mbar and voltage 5–10 kV.

3. EXPERIMENTAL RESULTS AND DISCUSSION

3.1. Images of LSCM and Fluorescence Spectroscopy

3.1.1. Effect of Gold Nanoparticles on the Fluorescence of NB Bound to Cationic Surfactants

The evaporation of the coated films prepared from the solutions No. 1 and 2 induced, respectively, a precipitation of aggregated cationic surfactants complexes with and without GNPs. Before the drying of the coated film (solution No. 2), NB as hydrophobic probe moves closely to the hydrophobic part of cationic surfactants [17]. As depicted in Figure 3, the fluorescence spectra of the untreated GNPs precipitation recorded with the lambda scan mode across the (530–775) nm range under an excitation of 514 nm shows a maximum at the wavelength of 565.59 nm (grey curve, Figure 3). Under the same excitation, the fluorescence spectra of treated GNPs precipitation recorded with the lambda scan mode across the same range is broadened and slightly redshifted compared to the fluorescence spectra in the absence of GNPs (orange curve, Figure 3). For the sample treated with GNPs, the evaporation of the solution induced their aggregation which was expected since that the dipole-dipole interaction between these nanoparticles is stronger when the distance between them decreases. The SPR of GNPs presents a broadening and red shift when they aggregate [22]. Moreover, it seems that NB entrapment inside the hydrophobic part of the bilayer polymeric film covering aggregated GNPs is responsible of a FRET between NB (acceptor) and aggregated GNPs (donor) which explains the fluorescence enhancement in the (565.59–706.59) nm range (Figure 3). When a fluorophore is quite close to a plasmonic particle, for instance within 10 nm, the electrons of the fluorophore are influenced by the plasmon field changing their excitation/emission properties. The interaction results in a quenching or enhancement of the fluorescence [23,24], which improves the performance of the optical response of the fluorescent contrast agents [25,26,27].

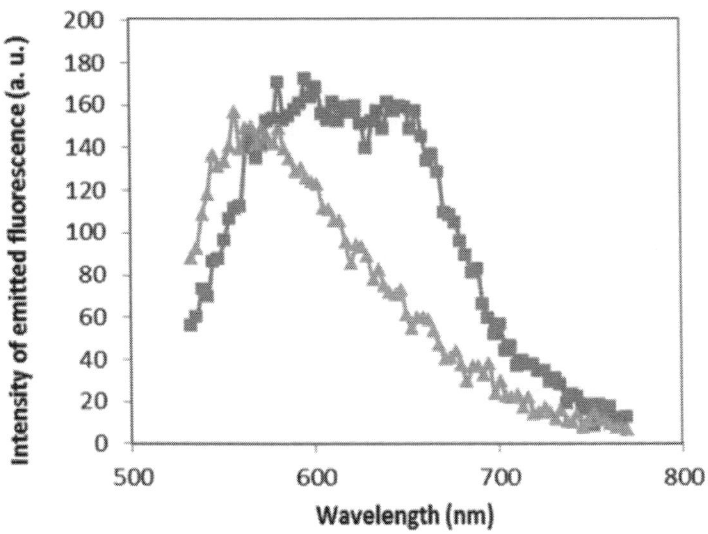

Figure 3. Effect of gold nanoparticles on the fluorescence spectra of NB bound to cationic surfactants complexes (grey and orange curves are recorded respectively from precipitations of aggregated cationic surfactants complexes without and with GNPs). Emission obtained from 514 nm excitation (lambda scan mode from a confocal microscope) and detected in the (530–775) nm range.

Thus, the FRET process between aggregated GNPs and NB induced a broadening of the fluorescence spectra of NB bound to cationic surfactants complexes.

Moreover, the slightly redshift of the fluorescence spectra of NB bound to cationic surfactant complexes in the presence of GNPs can be attributed to the SPR redshift of GNPs after their aggregation.

3.1.2. Effect of DNA on GNPs Aggregation

In the previous paragraph we showed that the broadening and the redshift of the fluorescence spectra of NB bound to cationic surfactants complexes was an indicator of GNPs aggregation.

In the following paragraph, the effect of DNA on GNPs aggregation size will be discussed.

As the DNA is concentrated (>1 mg/mL; solution No. 3), the molecules were precipitated into a liquid crystalline phase after the drying of the solution deposited via drop-coating on the glass slide [28]. Figure 4 shows emitted fluorescence from NB bound to DNA molecules organized into the flower like texture which is a highly ordered liquid crystalline form of DNA [29], detected in the (550–750) nm range under the wavelength of 514 nm. As depicted in Figure 5, the fluorescence spectra of NB bound to DNA molecules organized into liquid crystalline phase recorded with lambda scan mode from point marked as A is redshifted compared to the fluorescence spectral of NB bound to cationic surfactants complexes (Figure 3). This shift is explained by the sensitivity of NB to the polarity of the environment. The hydrophobic part of cationic surfactants, where NB is placed is non-polar, whereas in the case of DNA stained with NB, there is an amount of this probe which is bound to the polar negatively-charged phosphate backbone.

Figure 6 shows the emitted fluorescence, detected in the (550–750) nm ranges, from precipitated dried DNA/GNPs/NB solution under the excitation wavelength of 514 nm. A strong emitted light enhancement (green and yellow regions in Figure 6) is detected in the (550–560) nm range compared to the untreated sample with GNPs (Figure 4 and Figure 5). The emitted light detected in this range is attributed to an amount of NB entrapped inside the hydrophobic part of cationic surfactants stabilizing the GNPs suspension in citrate buffer. Moreover, as shown inside the region marked with the green color in Figure 6C, brighter spots are detected which suggested that GNPs are aggregated into spherical clusters. When NB are placed at proximity to the GNPs surface, a fluorescence enhancement is induced due to the FRET process. Additionally, GNPs plasmonic fields couple together when they are in close proximity [16]. For denser GNPs clusters, the plasmonic field coupling is increased, resulting in significantly higher absorption [30], which induced a higher FRET and brighter signal.

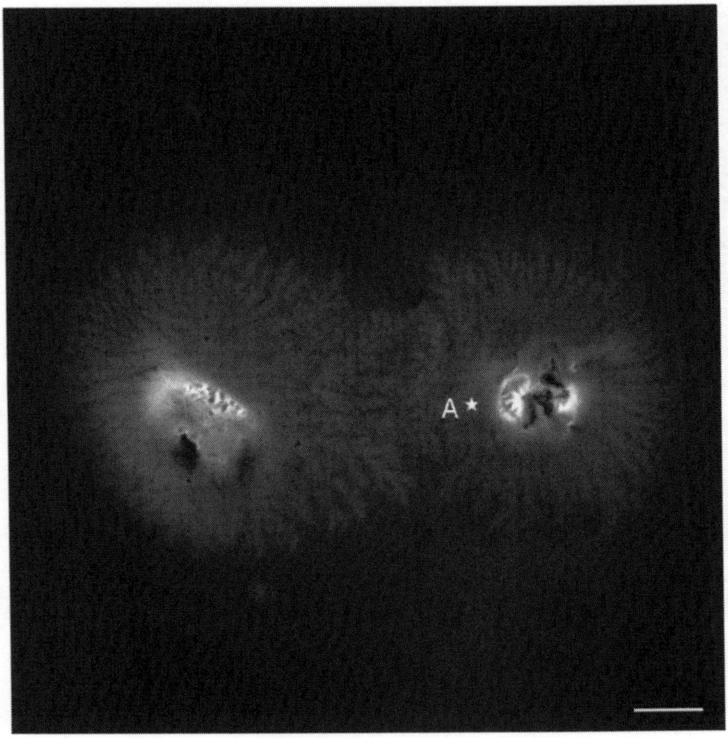

Figure 4. LSCM image of precipitated DNA (evaporated solution No. 3). The 514 nm was used as the pump beam and the fluorescent emission was detected in the (550–650) nm and (650–750) nm ranges. The scale bar donates 40 µm.

As shown in Figure 7, the fluorescence spectra recorded at the point marked as in Figure 6B is the same in point marked as A in Figure 4, whereas the fluorescence spectra recorded at the point marked as b in Figure 6B is broadened and blueshifted (Figure 7). As explained in the previous paragraph, the broadening of the fluorescence spectra is an indicator of GNPs aggregation. When GNPs are aggregated in the presence of DNA, an amount of NB inside the hydrophobic part of cationic surfactant bilayer film surrounding the GNPs is responsible of the blueshift.

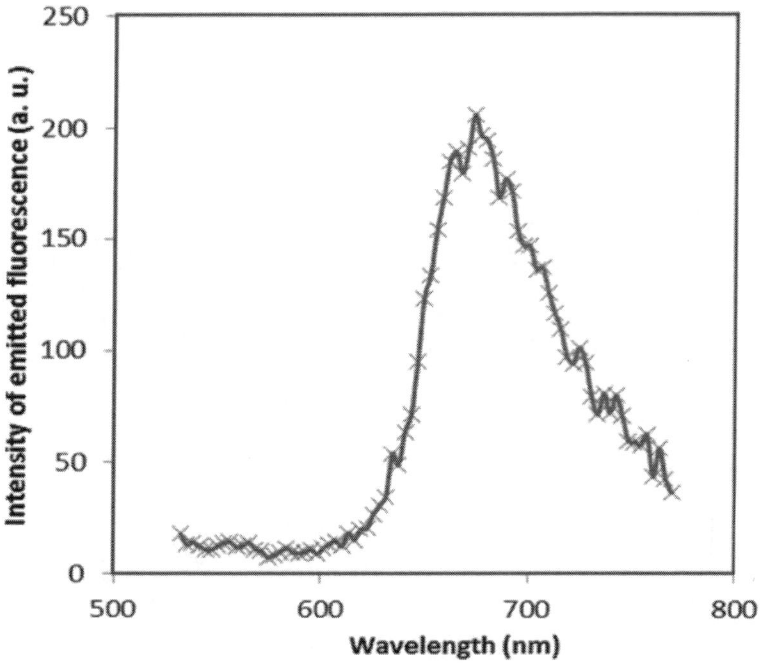

Figure 5. Fluorescence spectra of NB bound to DNA molecules organized into liquid crystalline phase recorded with lambda scan mode from the point marked as A (Figure 4). Emission obtained from 514 nm.

The fluorescence spectra recorded from aggregated GNPs in the presence of DNA (Figure 7 blue curve) is slightly redshifted compared to the fluorescence spectra recorded from aggregated GNPs without DNA (Figure 3 orange curve).

This redshift seems to be attributed to a growth of GNPs aggregation size induced by DNA compared to the spontaneous smaller GNPs aggregation induced by dipole-dipole interaction.

3.2. Images of ESEM

After their observations with confocal microscopy, the two precipitations of GNPs stabilized suspension in citrate buffer with and without DNA were stored at 4 °C for 24 h before ESEM observations. Figure 8A shows that for the sample untreated with DNA, GNPs are aggregated into some clusters. As previously discussed, this self-assembly is due to a dipole-dipole interaction between nanoparticles. In the presence of DNA, the GNPs aggregation size increased. GNPs are aggregated into denser three-dimensional clusters (Figure 8B). Moreover, for a higher magnification (100 nm scale bar), GNPs are assembled in 3-D structures following the helix morphology of DNA (Figure 8C). This aggregation growth is due to electrostatic and hydrophobic interactions between positive charged hydrophobic head groups of the bilayer cationic surfactants film around GNPs and negatively-charged DNA phosphate backbone [9,31].

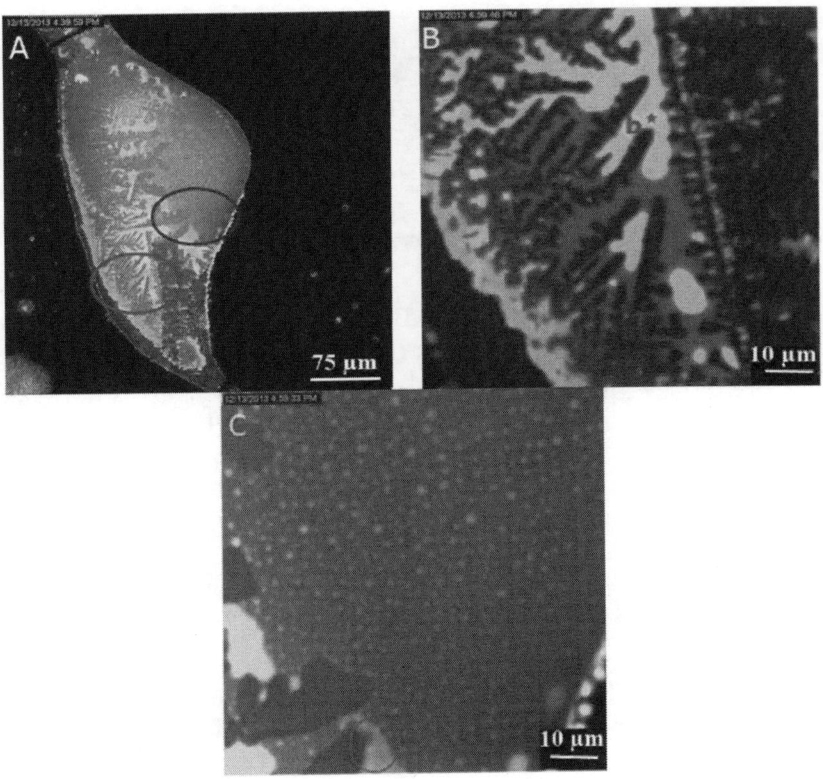

Figure 6. LSCM images of precipitated DNA/GNPs/solution. (**B**) and (**C**) are images of higher magnification at respectively the red and blue markets location in image (**A**). The 514 nm was used as pump beam. Light detected in (550–650) nm is green (Figure 6A,B) or yellow (Figure 6C) and light detected in (650–750) nm is red (Figure 6B,C) or grey (Figure 6A). The representation colors have been chosen to enhance the contrast.

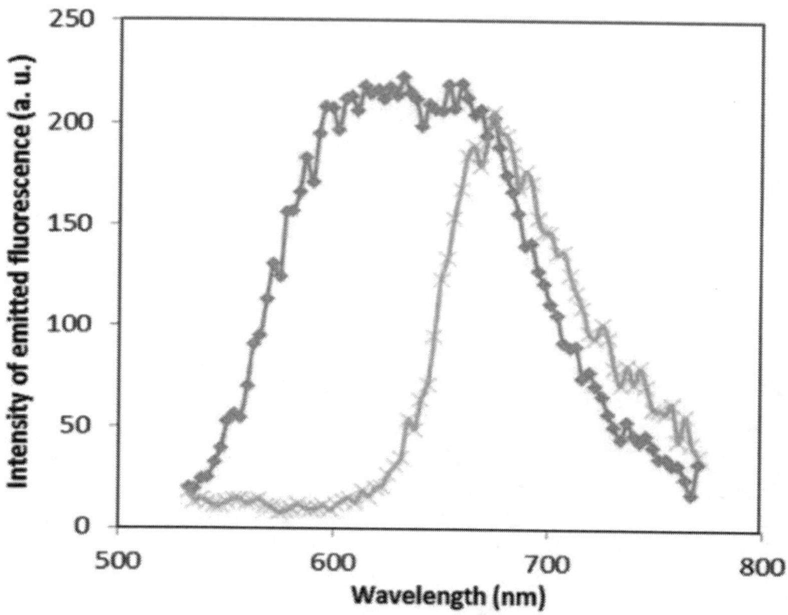

Figure 7. Fluorescence spectra of NB bound to precipitated DNA in the absence (yellow) and in the presence (blue) of GNPs aggregation recorded with lambda scan mode in respectively points marked as a and b in Figure 6B. The emission is obtained under an excitation of 514 nm and it is detected in the (530–775) nm range.

4. CONCLUSIONS

In this paper, FRET between GNPs (donors) and dyes (acceptors) is used for the detection of GNPs aggregation through LSCM. Moreover, DNA form Salmon tests was tested for the GNPs aggregation growth. The 514 nm was used as pump beam to irradiate the sample, since this wavelength is within the SPR of GNPs and the excitation bandwidth of the fluorescence dye (NB).

Thus, the fluorescence was manipulated with the SPR through the FRET process. The fluorescence spectra and peak are, respectively, broadened and shifted, which allowed the detection of the GNPs aggregation. In the presence of DNA, the fluorescence spectrum is more redshifted due to the size increase of GNPs aggregation induced by DNA. This result is confirmed with ESEM images which also showed a GNPs assembly into denser clusters in the presence of DNA. Moreover, for a higher ESEM image magnification, GNPs are assembled in 3-D structures following the helical morphology of DNA. DNA induced GNPs aggregation growth due to the electrostatic and hydrophobic interactions between positively-charged head groups of cationic surfactants coating GNPs and negatively-charged DNA phosphate backbone. Furthermore, these results may be helpful in the world of nanobiotechnology (gene delivery, cancer therapy, etc) mainly that the DNA-binding NB molecules are not toxic [17,18] and aggregated GNPs were usually detected via light scattering techniques, which can present problems since several structures within biological tissues also scatter light strongly. Thus, FRET between GNPs and dyes detected through LSCM result to be ideal for further bio-applications to discriminate aggregated GNPs and complex structures of tissue and cells.

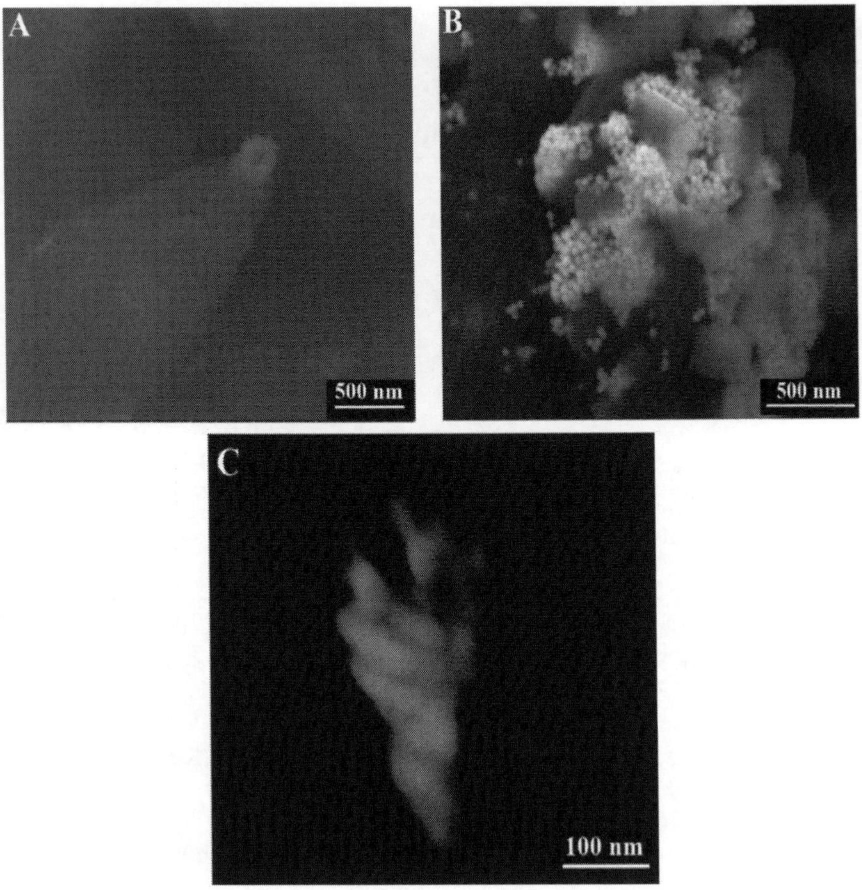

Figure 8. ESEM images of precipitates of evaporated; (**A**) GNPs-stabilized suspension in citrate buffer; (**B**,**C**) DNA/GNPs solutions.

ACKNOWLEDGMENTS

This work was supported by the grant LKE4CC_008 in the framework of PRIN 2010-2011. The Authors are grateful to Giovanni Desiderio for the ESEM images and Fernanda Annesi for samples preparation.

AUTHOR CONTRIBUTIONS

Riccardo Barberi and Giovani Carbone conceived and designed the experiments; Ramla Gary performed the experiments; Ramla Gary and Gia Petriashvili analyzed the data; Maria Penelope De Santo contributed reagents/materials/analysis tools; Ramla Gary wrote the paper.

CONFLICTS OF INTEREST

The authors declare no conflict of interest.

REFERENCES

1. Mirkin, C.A.; Letsinger, R.L.; Mucic, R.C.; Stohoff, J.J. A DNA-based method for rationally assembling nanoparticles into macroscopic materials. Nature **1996**, 382, 607–609.

2. Rahban, M.; Divsalar, A.; Saboury, A.A.; Golestani, A. Nanotoxicity and Spectroscopy Studies of Silver Nanoparticle: Calf Thymus DNA and K562 as Targets. J. Phys. Chem. C **2010**, 114, 5798–5803.

3. Basu, S.; Jana, S.; Pande, S.; Pal, T. Interaction of DNA bases with silver nanoparticles: Assembly quantified through SPRS and SERS. J. Colloid Interface Sci. **2008**, 321, 288–293.

4. Niemeyer, M.; Mirkin, C.A. NanoBiotechnology: Concepts, Methods and Applications; Wiley-VCH: Weinheim, Germany, 2004.

5. Storhoff, J.J.; Mirkin, C.A. Programmed materials synthesis with DNA. Chem. Rev. **1999**, 99, 1849–1862.

6. Niemeyer, C.M. Nanopartikel, Proteine und Nucleinsäuren: Die Biotechnologie Begegnet den Materialwissenschaften. Angew. Chem. **2001**, 113, 4254–4287. (In Germany).

7. Braun, E.; Eichen, Y.; Sivan, U.; Ben-Yoseph, G. DNA-templated assembly and electrode attachment of a conducting silver wire. Nature **1998**, 391, 775–778.

8. Single Stranded DNA Induced Assembly of Gold Nanoparticles. Available online: http://dspace.mit.edu/handle/1721.1/3949 (accessed on 31 July 2015).

9. Ganech, K.N.; Sastry, M. DNA-amine interactions: From monolayers to nanoparticles. J. Indian Inst. Sci. **2002**, 82, 105–112.

10. Nakao, H.; Shiigi, H.; Yamamoto, Y.; Tokonami, S.; Nagaoka, T.; Sugiyama, S.; Ohtani, T. Highly Ordered Assemblies of Au Nanoparticles Organized on DNA. Nano Lett. **2003**, 3, 1391–1394.

11. Lee, J.S.; Han, M.S.; Mirkin, C.A. Colorimetric Detection of Mercuric Ion (Hg^{2+}) in Aqueous Media Using DNA-Functionalized Gold Nanoparticles. Angew. Chem. **2007**, 119, 4171–4174.

12. Kanayama, N.; Takarada, T.; Maeda, M. Rapid naked-eye detection of mercury ions based on non-crosslinking aggregation of double-stranded DNA-carrying gold nanoparticles. Chem. Commun. **2011**, 47, 2077–2086.

13. De la Rica, R.; Stevens, M. Plasmonic ELISA for the ultrasensitive detection of disease biomarkers with the naked eye. Nat. Nanotechnol. **2012**, 7, 821–824.

14. Tsai, S.; Chen, Y.; Liaw, J. Compound Cellular Imaging of Laser Scanning Confocal Microscopy by Using Gold Nanoparticles and Dyes. Sensors **2008**, 8, 2306–2316.

15. Durr, N.J.; Larson, T.; Smith, D.K.; Korgel, B.A.; Sokolov, K.; Ben-Yarkar, A. Two-Photon Luminescence Imaging of Cancer Cells Using Molecularly Targeted Gold Nanorods. Nano Lett. **2007**, 7, 941–945.

16. Aioub, M.; Kang, B.; Mackey, M.A.; El-Sayed, M.A. Biological Targeting of Plasmonic Nanoparticles Improves Cellular Imaging via the Enhanced Scattering in the Aggregates Formed. J. Phys. Chem. Lett. **2014**, 5, 2555–2561.

17. Mitra, R.K.; Sinha, S.S.; Maiti, S.; Pal, S.K. Interactions of Nile Blue with Micelles, Reverse Micelles and a Genomic DNA. J. Fluoresc. **2008**, 18, 423–432.

18. Mitra, R.K.; Sinha, S.S.; Maiti, S.; Pal, S.K. Sequence Dependent Ultrafast Electron Transfer of Nile Blue in Oligonucleotides. J. Fluoresc. **2009**, 19, 353–361.

19. Jim, J.; Lee, N.; Le, E.; Yoon, S. Surface Modification of Citrate-Capped Gold Nanoparticles Using CTAB Micelles. Bull. Korean Chem. Soc. **2014**, 35, 2567–2569.

20. Jana, N.R.; Gearheart, L.; Murphy, C.J. Seeding Growth for Size Control of 5–40 nm Diameter Gold Nanoparticles. Langmuir **2001**, 17, 6782–6786.

21. Oregon Medical Laser Center. Nile Blue. Available online: http://omlc.org/spectra/PhotochemCAD/html/068.html (accessed on 23 June 2015).

22. Gao, Q.; Zheng, Y.; Song, C.; Lu, L.; Tian, X.; Xu, A. Selective and sensitive colorimetric detection of copper ions based on anti-aggregation of the glutathioneinduced aggregated gold nanoparticles and its application for determining sulfide anions. RSC Adv. **2013**, 3, 21424–21430.

23. Kang, K.A.; Wang, J.; Jasinski, J.B.; Achilefu, S. Fluorescence Manipulation by Gold Nanoparticles: From Complete Quenching to Extensive Enhancement. J. Nanobiotechnol. **2011**, 9, 1–13.

24. Schneider, G.; Decher, G.; Nerambourg, N.; Praho, R.; Werts, M.H.V.; Blanchard-Desce, M. Distance-dependent fluorescence quenching on gold nanoparticles ensheathed with layer-by-layer assembled polyelectrolytes. Nano Lett. **2006**, 6, 530–536.

25. My, N.; Wc, L. Fluorescence enhancements of fiber-optic biosensor with metallic nanoparticles. Opt. Expr. **2009**, 17, 5867–5878.

26. Kang, K.A.; Hong, B. Biocompatible nano-metal particle fluorescence enhancers. Crit. Rev. Eukaryot. Gene Expr. **2006**, 16, 45–60.

27. Gary, R.; Amelio, D.; Garofalo, F.; Petriashvili, G.; de Santo, M.P.; Kwong Ip, Y.; Barberi, R. Endothelial-like nitric oxide synthase immunolocalization by using gold nanoparticles and dyes. Biomed. Opt. Expr. **2015**, 6, 4738–4748.

28. Livolant, F.; Leforestier, A. Condensed phases of DNA: Structures and phase transitions. Prog. Polymer. Sci. **1996**, 21, 1115–1164.

29. Saminathan, M.; Thomas, T.; Shirahata, A.; Pillai, C.K.S.; Thomas, T.J. Polyamine structural effects on the induction and stabilization of liquid crystalline DNA: Potential applications to DNA packaging, gene therapy and polyamine therapeutics. Nucleic Acids Res. **2002**, 30, 3722–3731. [

30. Thanh, N.T.K.; Vernhet, A.; Rosenzweig, Z. Gold Nanoparticles in Bioanalytical Assays and Sensors. Springer Ser. Chem. Sens. Biosens. **2005**, 3, 261–277.

31. Sun, L.; Sun, Y.; Zhang, F.X.; Yang, T.; Guo, C.; Liu, Z.; Li, Z. Atomic force microscopy and surface-enhanced Raman scattering detection of DNA based on DNA–nanoparticle complexes. Nanotechnology **2009**, 20, 125502–125514.

CHAPTER 6

Thermal Diffusivity and Thermal Conductivity of Epoxy-Based Nanocomposites by the Laser Flash and Differential Scanning Calorimetry Techniques

Jasper Chiguma[1], Edwin Johnson[2], Preyah Shah[2], Natalya Gornopolskaya[2], Wayne E. Jones Jr.[2*]

[1]Materials Science and Engineering Program, Binghamton University, Binghamton, USA; [2]Department of Chemistry, Binghamton University, Binghamton, USA.

ABSTRACT

Nanocomposites were fabricated by using a commercial two part epoxy as a matrix and multiwalled carbon nanotubes, graphite fibers and boron nitride platelets as filler materials. Multiwalled carbon nanotubes (MWCNTs) that were produced by chemical vapor deposition were found to produce nanocomposites with better thermal diffusivity and thermal conductivity than the MWCNTs that were produced by the combustion method. Compared to the MWCNTs produced by both methods and graphite fibers, boron nitride produced nanocomposites with the highest thermal conductivity. Specific heat capacity was measured by using differential scanning calorimetry and thermal diffusivity was measured by using the laser flash.

KEYWORDS

Nanocomposites; Thermal Diffusivity; Thermal Conductivity; Heat Capacity; Laser Flash; Differential Scanning Calorimetry

1. INTRODUCTION

One of the major challenges that are encountered in the packaging of semiconductor electronic devices is the progressive increase in the power level of operating devices. Balancing the reduction of semiconductor device feature sizes with reliable operation continues to be a challenge. The effective transfer of heat from an intergrated circuit and its heat spreader to a heat sink is a vital step in solving the challenge. Thermal management within the area of the chip calls for addressing thermal impedance in the following areas: junction-to-case, case-to-heat sink and heat sink-to-ambient [1].

Nanocomposites with superior thermal transport properties have great application potential in microelectronics as thermal interface materials (TIMs). A polymer matrix and a thermally conducting organic or inorganic filler with nanoscale dimensions are the two components that constitute a nanocomposite that is suitable for thermal interface applications. Some organic fillers that have been used as fillers are carbon nanotubes, graphite fibers and more recently graphene. Usually, materials such as Al_2O_3, AlN and BN are used as inorganic fillers [2,3].

Epoxy resins are among some of the polymer materials that are resistant to adverse conditions and are characterized by good stiffness, strength, stability and chemical resistance and adhere well to embedded fillers [4]. As a result, they have been used in a lot of electronics such as resistors, inductors, capacitors, lasers, low loss dielectric and waveguide where toughness and resistance to high temperatures are required. With the increase in demand for affordable interfacial materials that are capable of dissipating heat effectively in electronic systems, nanomaterials like carbon nanotubes have been incorporated in epoxies and other suitable polymers such as polyimides to form nanocomposites [5]. The choice of polymer matrix is greatly influenced by the anticipated area of application. For instance, high temperature applications in spaceships would require the use of polyimides and epoxies [6].

The filler materials are selected based on their properties which they are expected to impart on the polymer matrix. Besides carbon nanotubes, other materials that have been used as fillers include nanotubes and nanoparticles of metal as well graphite fibers among others [7].

Melt mixing, solution casting and in-situ polymerizetion are the three predominant ways of preparing nanocomposites[8]. Solution casting is mostly used because it does not require any specialized equipment to melt polymers some of which melt at very high temperatures. A major issue associated with the fabrication of nanocomposites is the dispersion of nanofillers in the polymer matrix [9,10]. While they have high thermal conductivity, estimated at 6000 $Wm^{-1}{\cdot}K^{-1}$ and carrier mobility estimated to be 10^5 $cm^2{\cdot}V^{-1}{\cdot}s^{-1}$, carbon nanotubes have high interfacial energy as a result of their high aspect ratio which causes them to agglomerate and as a result make it difficult to disperse them in solvents [11]. Bath sonication, tip sonication, functionalization and use of surfactants are some of the methods that are currently being used to disperse carbon nanotubes in order to overcome the problem of agglomeration [12]. With the dispersion of carbon nanotubes it is anticipated that a continuous percolation pathway through which current can flow is formed.

A percolation network or continuous path from one end of a material to the other is formed when the number of nanotubes per unit volume of polymer exceeds a certain threshold value [13]. However, too much dispersion has the disadvantage of having isolated carbon nanotube strands insulated with the polymer matrix resulting in higher resistance. It has also been found that at higher concentrations of carbon nanotubes, the aggregated sample tend to conduct better [14]. Ideally, electrical percolation at the smallest possible loading of carbon nanotubes is desired, yet this can be made non-desirable by polymer encapsulation. Despite all these challenges, due to the many envisioned applications of carbon nanotubepolymer nanocomposites, researchers in many branches of materials science continue to find ways to enhance their mechanical, electrical and electronic properties.

Table 1 shows the materials that were used as fillers in the preparation of epoxy nanocomposites. All these materials were obtained courtesy of Nanodynamics Inc.

Buffalo. Except for the CVD produced MWCNTs, all the carbon nanotubes were products of Nanodynamics which were produced by the combustion of ethylene in air with the use of a nickel catalyst.

There are four major methods that have been used to measure thermal conductivity, namely the Hot wire method, Laser Flash, Thermal conductivity from the Dynamic Response of the DSC and the Scanning Thermal Microscopy [15-19]. In this work we make use of the flash method. In the thermal flash method, the front face of a small wall-shaped sample receives a pulse of radiant energy coming from a laser or a flash lamp.

Figure 1 illustrates how the thermal diffusivity value is computed from the resulting temperature response on the opposite (rear) face of the sample. The speed with which

$$\lambda = \alpha\rho C_p \tag{1}$$

heat propagates through a material is known as thermal diffusivity. It is related to thermal conductivity by equation 1, where, λ is thermal conductivity in $Wm^{-1}\cdot K^{-1}$, α is thermal diffusivity in $m^2\cdot s^{-1}$, ρ is density in $kg\cdot m^{-3}$ and C_p is specific heat capacity in $J\cdot kg^{-1}\cdot K^{-1}$.

Table 1. Fillers for nanocomposites.

Filler	Code	Properties
1) MWCNTs	(N7)	CVD produced (pristine)
2) MWCNTs	(AC)	Combustion produced (Acid treated)
3) MWCNTs	(AL-ALD)	Combustion produced (ALD coated)
4) Graphite Fibers	(GF)	Pristine
5) Boron Nitride	(BN)	Uncoated
6) Boron Nitride	(BN10)	Alumina ALD coated (10 cyles)
7) Boron Nitride	(BN25)	Alumina ALD coated (25 cyles)

Thermal conductivity is the time rate of steady heat flow through unit thickness of an infinite slab of homogeneous material in a direction perpendicular to the surface, induced by unit temperature difference. It is the property that determines the working temperature levels of a material. The amount of heat required to raise the temperature of a given substance or material by 1˚C or 1 K, is the heat capacity of the material. Specific heat capacity is a thermodynamic quantity that determines the amount of heat necessary to raise by one degree the temperature of a unit mass of material [20].

In order to calculate the thermal conductivity, α has to be determined as well as the C_p. Equation (2) represents the Fourier series solution that represents the temperature rise of the backside of the sample.

$$T(L,t)=\frac{Q}{\rho CpL}\left[1+2\sum_{n=1}^{\infty}(-1)^n\exp\left(\frac{-n^2\pi^2}{L^2}\alpha t\right)\right] \quad (2)$$

Q, ρ, C_p and L are the energy of the laser pulse per unit area, density of the sample, specific heat capacity of the sample and thickness of the sample respectively. In order to normalize Equation (2) in terms of the maximum backside temperature ($T_{max}(L)$), Equations (3) and (4) have been defined.

$$V(L,t)=T(L,t)/T_{max}(L) \quad (3)$$

$$\omega=\pi^2\alpha t/L^2 \quad (4)$$

Equations (3) and (4) lead to the Fourier series Equation (5)

$$V(L,t)=1+2\sum_{N=1}^{\infty}(-1)^n\exp(-n^2\omega). \quad (5)$$

when V(L,t) is 0.5, that is 50% of the peak value, ω is equal to 1.388 and the expression of α will be Equation (6).

$$\alpha=1.388L^2/(\pi^2 t_{1/2}). \quad (6)$$

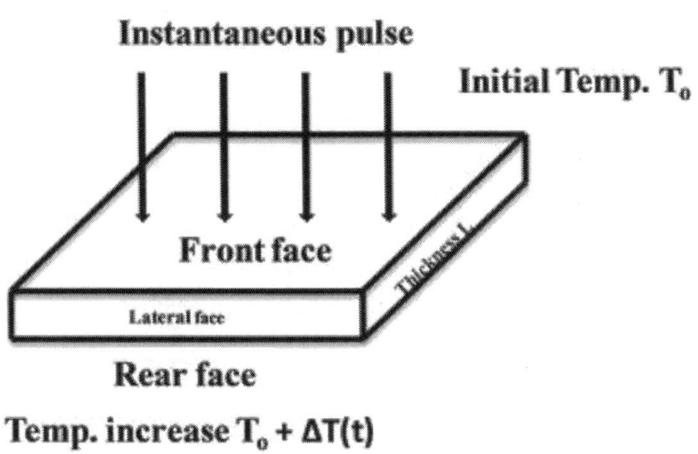

Figure 1. Schematic of the flash method.

In Equation (6), $t_{1/2}$ is the time required for the backside temperature to reach half of the maximum temperature rise. This is the equation that is used to calculate thermal diffusivity by measuring $t_{1/2}$ when the dimensionless temperature V(L,t) is equal to 0.5. Figure 2 shows the relationship between V(L,t) and ω [21].

The specific heat capacity, C_p, is related to the amount of heat, the substance in question and temperature by the Equation (7):

$$Q = mC_p\Delta T. \tag{7}$$

where Q is the amount heat, m is the mass of substance and ΔT is the change in temperature. The thermal analysis tool that is used to determine specific heat capacity is the DSC. It measures the temperature and heat flow associated with important transitions in materials as a function of time and temperature. Glass transition temperature, melting point, latent heat of melting, latent heat of crystallization, endothermic and exothermic natures of transition, degree of crystallinity, phase changes are other measurements that can be obtained by DSC besides specific heat capacity [22].

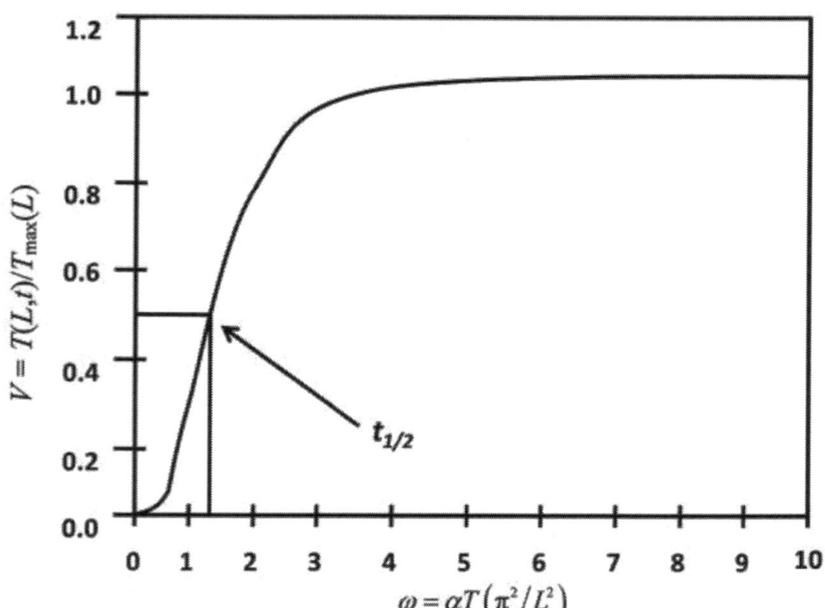

Figure 2. Relation between V(L,t) and ω.

While the method is based on an ideal method which assumes that the sample is homogeneous and isotropic, and the thermophysical properties and the density are uniform, constant and invariant with temperature within the experimental conditions, among other assumptions, this is not always the case with composites. The extension of the flash method to composites assumes that the sample thickness is much larger than the fiber diameter, the fiber volume fraction is as large as possible, and the matrix is in perfect thermal contact, that is, the interfacial conductance is large [23,24].

Given the nanoscale diameter of our nanotubes and the microscale dimensions of the carbon fibers and thegraphite fibers we have the confidence that our composites meet the requirements for thermal diffusivity measurements using the flash method.

2. EXPERIMENTAL

A commercial epoxy resin (Epon Resin 815C) and a curing agent (Epikure 3140 Curing Agent) were obtained from Miller Stephenson Chemicals Co. Inc. and Methyl Ethyl Ketone (MEK) was obtained from Aldrich Chemicals Company. All chemical were used as received. MWCNTs in 5 ml of MEK were bath sonicated for 1hr to achieve dispersion followed by the addition of 2 g of Epon Resin 815C and further sonication for 30 minutes to achieve mixing. The suspensions were then poured into Teflon mold cups and left overnight to evaporate the solvent. This was followed by the addition of 2 g of the Epikure 3140 Curing Agent. After thorough mixing, the nanocomposites were degassed under vacuum for an hour and then left to cure at room temperature for 72 hours. Samples of dimensions 12.7 mm × 12.7 mm × 0.134 mm were cut using a diamond saw and polished to achieve smooth surfaces. Thermal diffusivity data were collected using a FlashLine™2000 Thermal Properties Analyzer from Anter Corporation. This instrument has the following attributes: lower cost-bench-top system, xenon flash, operating temperature from RT to 330°C, loading capacity of four specimen (30 mm diameter), specific heat capacity determination, thermal conductivity measurement, ASTM E-1461. Samples for thermal diffusivity measurements were coated with a thin layer of graphite to increase emissivity and eliminate transparency. The volumes of the samples were determined from the length, width and thickness of the samples. The densities of the samples were then calculated from their mass and volume.

The specific heat capacity of each sample was determined using a Differential Scanning Calorimeter (DSC). The heating procedure involved a number of steps. Firstly, heating was done in the absence of a reference pan and a sample pan in order to achieve baseline correction. The pans are made of sapphire. Secondly, heating was done in the presence of empty reference and sample pans to measure background heat flow. Thirdly, the calibration constant k was calculated. Lastly, each weighed sample was run through the calorimeter to determine the heat flow at room temperature and calculate the corresponding C_p. The DSC used nitrogen as a purge gas and the mass flow set at 50 ml/min for the sample purge. The ramp rate was 3.000°C/min up to 50°C.

Using the outlined procedure, Epon Resin 815C nanocomposites were prepared with loadings of MWCNTs, Graphite fibers and Boron Nitride ranging from 0 - 0.5 wt%. In order to see whether there would be improvement in thermal properties at higher loadings, nanocomposites with MWCNTs and graphite fibers were also prepared with percent loadings of 0 - 25 wt%. In order to improve the chances of seeing the filler materials using Field Emission Scanning Electron Microscopy (FESEM) type Supra 55, the cast nanocomposites were fractured and the lateral side examined. The MWCNTs were analyzed using High Resolution Transmission Electron Microscopy (HRTEM).

3. RESULTS AND DISCUSSION

Figure 3 shows the surface morphology of the fillers. The micrographs were obtained using FESEM and HRTEM. While the structure of the fillers can be clearly seen, it was not possible to see them in the nanocomposites. In order to increase the chances of seeing the fillers in the nanocomposites, the nanocomposites were fractured. Figure 4 shows FESEM images of the fractured surfaces. Table 2 shows the diffusivity data that were collected for the epoxy nanocomposites at each percentage loading. The tabulated thermal diffusivity data at each percentage loading were obtained by averaging three thermal diffusivity readings that corresponded to three half-times and three laser shots. The laser flash gives thermal diffusivity data output that corresponds to different models that have been used to calculate thermal diffusivity. Among these models are those by Parker, Koski, Heckman, Cowan and Cowan and Taylor. The diffusivity data by all the models were very much in agreement. In this work, we selected the diffusivity data from the Parker model and plotted the average diffusivity from the three shots as a function of percentage loading.

The same procedure was followed to process the data that were collected for all the nanocomposites filled with the materials listed in Table 1. Figure 5 shows plots of the thermal diffusivities of the epoxy nanocomposites with all the carbon nanotubes as well as Graphite fibers at filler loadings of 0.1, 0.3 and 0.5 wt%. It can be seen in these plots how each filler material influences the thermal diffusivity of the epoxy matrix. In Figure 5, we see that within the same percentage loading the nanocomposite loaded with CVD produced MWCNTs achieves the highest thermal diffusivity of 1.87 × 10^{-3} $cm^2 \cdot s^{-1}$ at 0.3 wt% loading before dropping slightly to 1.73 × 10^{-3} $cm^2 \cdot s^{-1}$.

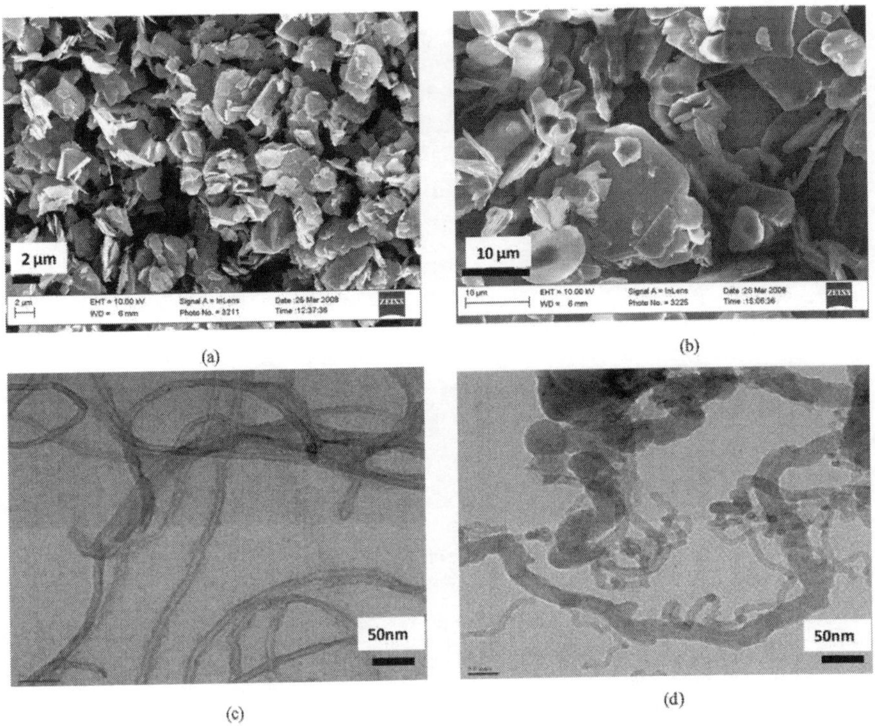

Figure 3. FESEM of (a) Graphite fibers (b) Boron Nitride Platelets (c) TEM, CVD MWCNTs (d) TEM, Acid treated MWCNTs.

The nanocomposite loaded with acid treated MWCNTs achieved a maximum diffusivity of 1.67 × 10^{-3} $cm^2 \cdot s^{-1}$ at 0.1 wt% loading, the highest of them all at that low loading. After that increase a continuous drop is observed. Just like the CVD produced MWCNTs filled nanocomposite, the ALD coated MWCNTs loaded nanocomposite showed a maximum in thermal diffusivity of 1.60 × 10^{-3} $cm^2 \cdot s^{-1}$ at 0.3 wt%. However, the thermal diffusivity value is lower than that of the CVD produced MWCNT filled nanocomposite at 0.3 wt%. The Graphite fiber loaded nanocomposite exhibits a gradual increase in thermal diffusivity and it reaches a maximum of 1.77 × 10^{-3} cm^2 × s^{-1} at 0.5 wt% loading. Figure 6 shows the plots of the specific heat capacities of each nanocomposite in the 18°C - 30°C range. The specific heat capacity in all cases increases systematically with temperature.

The data were obtained from DSC measurements. Figure 7 shows the plots of the specific heat capacities at each loading at 25°C for each nanocomposite. While the specific heat capacities of the other nanocomposites decreases at 0.5 wt% loading, the specific heat capacity of the CVD produced MWCNTs loaded nanocomposite continues to increase. From these specific heat capacity data, density of nanocomposite samples and their thermal diffusivity values, thermal conductivities for each sample at each loading were calculated. Table 3 shows the thermal conductivity data of the nanocomposites. Figure 8 shows the plots of the thermal conductivity data shown in Table 3. The CVD produced MWCNTs filled nanocomposite achieved the highest thermal conductivity. In order to find out what the thermal diffusivity trend would be with further filler loadings, nanocomposites were prepared with filler loadings of up to 25 wt%.

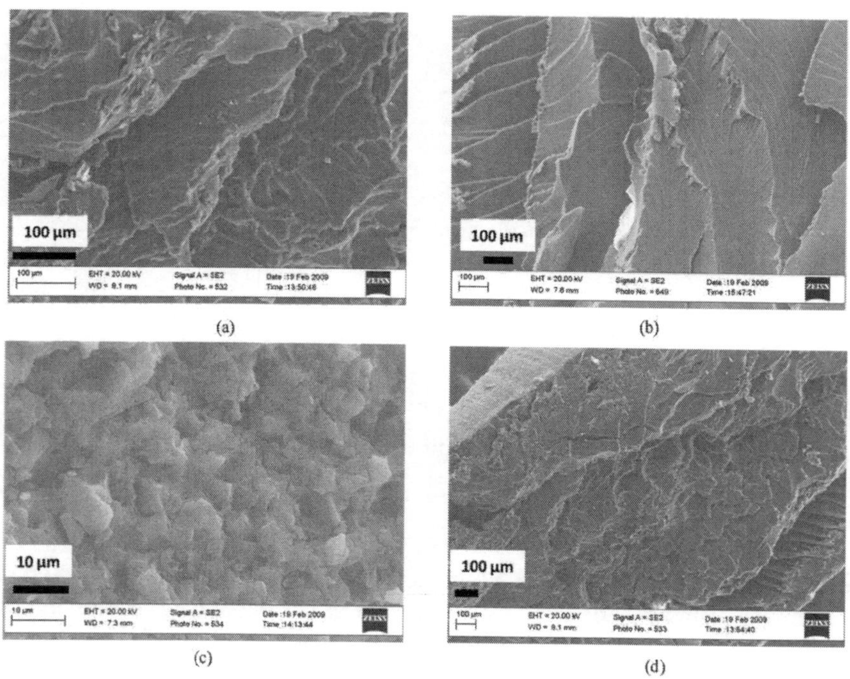

Figure 4. (a) Epoxy/MWCNTs (b) Epoxy/BN (c) Epoxy/Carbon Fibers (d) Fracture surface of nanocomposite.

Table 2. Diffusivity data of epoxy nanocomposites at each loading.

Loading (wt%)	α (Epoxy/AC) [$cm^2 \cdot s^{-1}$]	α (Epoxy/N7) [$cm^2 \cdot s^{-1}$]	α (Epoxy/GF) [$cm^2 \cdot s^{-1}$]	α (Epoxy/ALD) [$cm^2 \cdot s^{-1}$]
0.1	0.00167	0.00160	0.00150	0.00133
0.3	0.00147	0.00187	0.00143	0.00160
0.5	0.00157	0. 00173	0.00177	0.00146

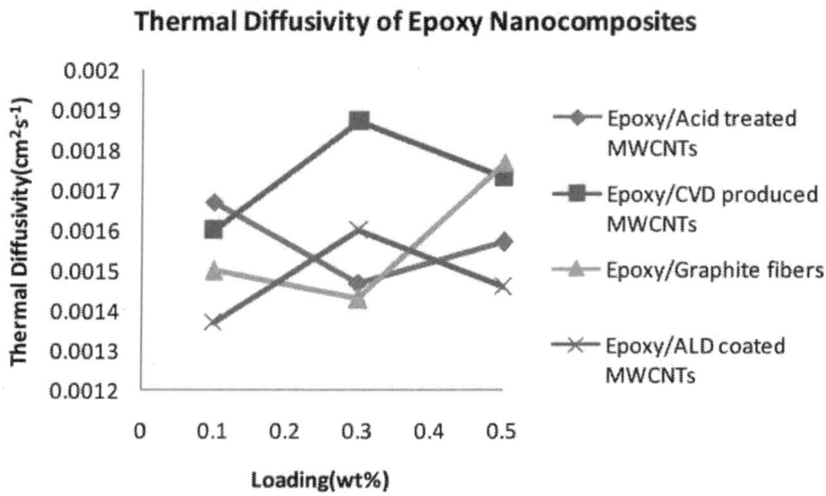

Figure 5. Plot of thermal diffusivity of epoxy nanocomposites as a function of percent loading of fillers.

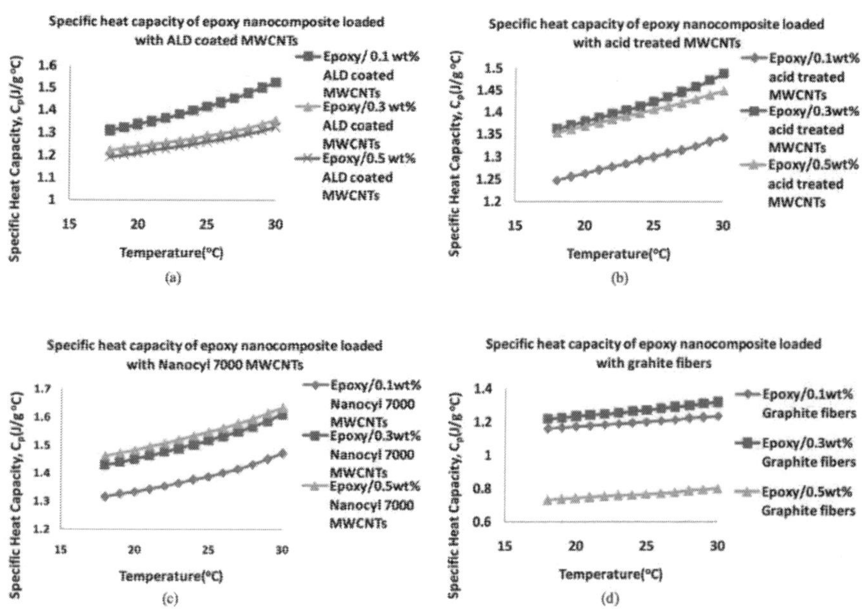

Figure 6. Plots of specific heat capacities of epoxy nanocomposites in the 18°C - 30°C temperature range filled with (a) ALD coated MWCNTs (b) Acid treated MWCNTs (c) CVD produced MWCNTs (d) Graphite fibers.

Nanocomposites were successfully prepared with all these fillers except the CVD produced MWCNTs because no good dispersion could be achieved at higher loadings. Table 4 shows the thermal diffusivity data that were obtained at 5, 15 and 25 wt% filler loading. Figure 9 shows the plots of the thermal diffusivity data. The plotted thermal diffusivity data show that the acid treated MWCNTs and the ALD coated MWCNTs loaded nanocomposites show maximum thermal diffusivity of 1.90×10^{-3} $cm^2 \cdot s^{-1}$ and 1.80×10^{-3} $cm^2 \cdot s^{-1}$ respectively at 15 wt% loading. These values are slightly higher than those that were obtained at lower loadings in the 0 - 0.5 wt% range. It seems that continued addition of these MWCNTs does not significantly increase the thermal diffusivity of the epoxy nanocomposite. Graphite fibers show an interesting trend in which the thermal diffusivity of the epoxy/ graphite fiber nanocomposite continues to increase with increasing loading of graphite fibers and reaching 3.80×10^{-3} $cm^2 \cdot s^{-1}$.

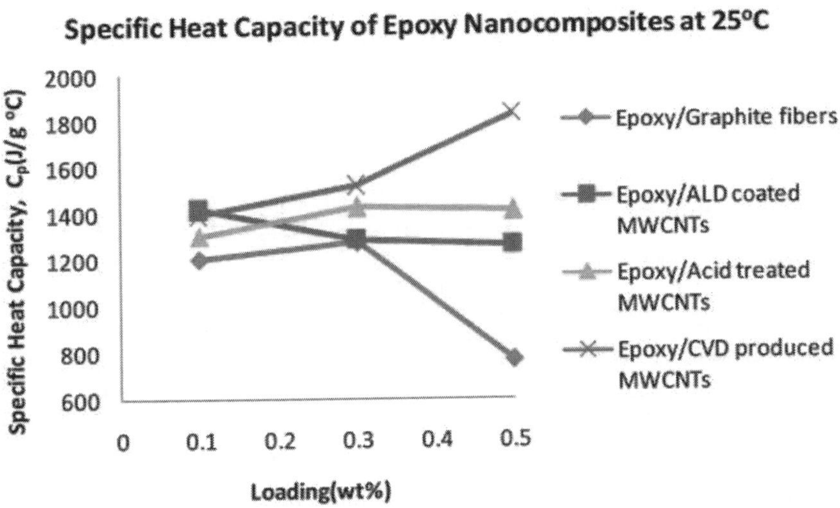

Figure 7. Plots of specific heat capacities of epoxy nanocompositesat 25°C as the filler loading increases.

Figure 10 shows the plots of the specific heat capacities in the 18°C - 30°C temperature range for the nanocomposites that were loaded at 5, 15 and 25 wt%. Generally, the specific heat capacity is found to increase with increase in temperature. Figure 11 shows plots of the specific heat capacities at 25°C. The plots show an increasing trend in the specific heat capacities of the nanocomposites for all of the nanocomposites. The graphite loaded nanocomposite shows the highest specific heat capacity of 4.20 $Jg^{-1} \cdot {}^{\circ}C^{-1}$ at 25 wt% loading.

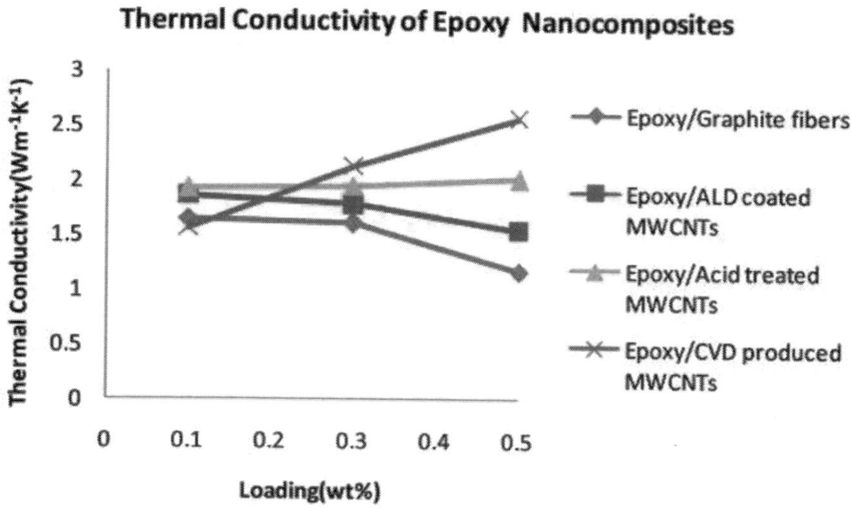

Figure 8. Plots of thermal conductivities of epoxy nanocomposites from 0.1 to 0.5 wt% loading

From the thermal diffusivity data, the sample density and specific heat capacity, the thermal conductivities of the nanocomposites were calculated. Table 5 summarizes the thermal conductivity results. Figure 12 shows plots of the thermal conductivity data that are tabulated in Table 5.

The plots reveal a superior performance by the graphite fiber filled nanocomposite, with thermal conductivity that increases systematically with percent loading to 16.25 $Wm^{-1}\cdot K^{-1}$. Behind this nanocomposite is the nanocomposite filled with ALD alumina coated MWCNTs, reaching a thermal conductivity of 12.30 $Wm^{-1}\cdot K^{-1}$.

This high thermal conductivity value for alumina coated carbon nanotubes shows that the coating has the effect of increasing thermal conductivity. The plot of the nanocomposite that was filled with uncoated carbon nanotubes shows much lower thermal conductivity compared to graphite fibers and alumina coated filled nanocomposites.

Table 6 is a summary of the thermal diffusivity of boron nitride filled nanocomposites. Figure 13 shows the thermal diffusivity plots of boron nitride loaded epoxy nanocomposites. The three types of boron nitride have different properties due to different treatments. One set of boron nitride is uncoated (BN uncoated) and the other two samples have been coated with alumina, one for 10 cycles (BN10) and the other for 25 cycles (BN25).

Table 3. Thermal conductivity data of epoxy nanocomposites.

Composite	α	C_p	λ
	$[m^2 \cdot s^{-1}]$	$[JKg^{-1} \cdot K^{-1}]$	$[Wm^{-1} \cdot K^{-1}]$
0.1 wt% CF	1.50×10^{-7}	1201.0	1.64
0.3 wt% CF	1.43×10^{-7}	1271.0	1.60
0.5 wt% CF	1.77×10^{-7}	767.7	1.17
0.1 wt% AL-ALD	1.37×10^{-7}	1413.0	1.85
0.3 wt% AL-ALD	1.60×10^{-7}	1284.0	1.78
0.5 wt% AL-ALD	1.46×10^{-7}	1258.0	1.53
0.1 wt% AC	1.67×10^{-7}	1298.0	1.91
0.3 wt% AC	1.47×10^{-7}	1423.0	1.94
0.5wt% AC	1.57×10^{-7}	1406.0	2.00
0.1 wt% NC7	1.60×10^{-7}	1387.0	1.55
0.3 wt% NC7	1.87×10^{-7}	1514.0	2.12
0.5 wt% NC7	1.73×10^{-7}	1825.0	2.56

Table 4. Diffusivity data of epoxy nanocomposites at each loading.

Loading	α (Epoxy/AC)	α (Epoxy/ALD)	α (Epoxy/GF)
(wt%)	$[cm^2 \cdot s^{-1}]$	$[cm^2 \cdot s^{-1}]$	$[cm^2 \cdot s^{-1}]$
5	0.00137	0.0015	0.00200
15	0.00190	0.0018	0.00267
25	0.00140	0.00177	0.00380

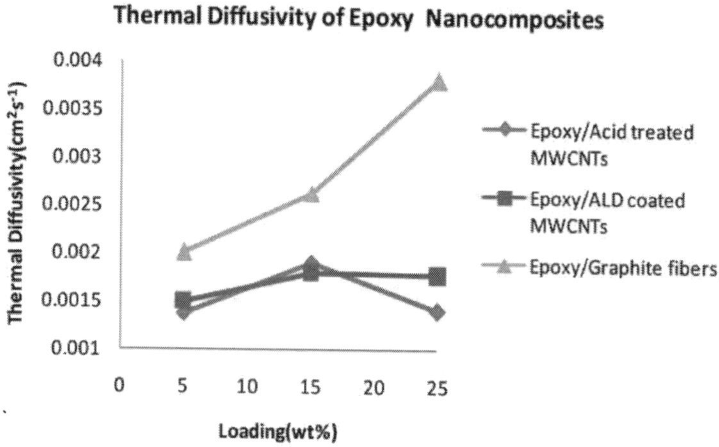

Figure 9. Plot of thermal diffusivity of epoxy nanocomposites as a function of percent loading of fillers (5 - 25 wt%).

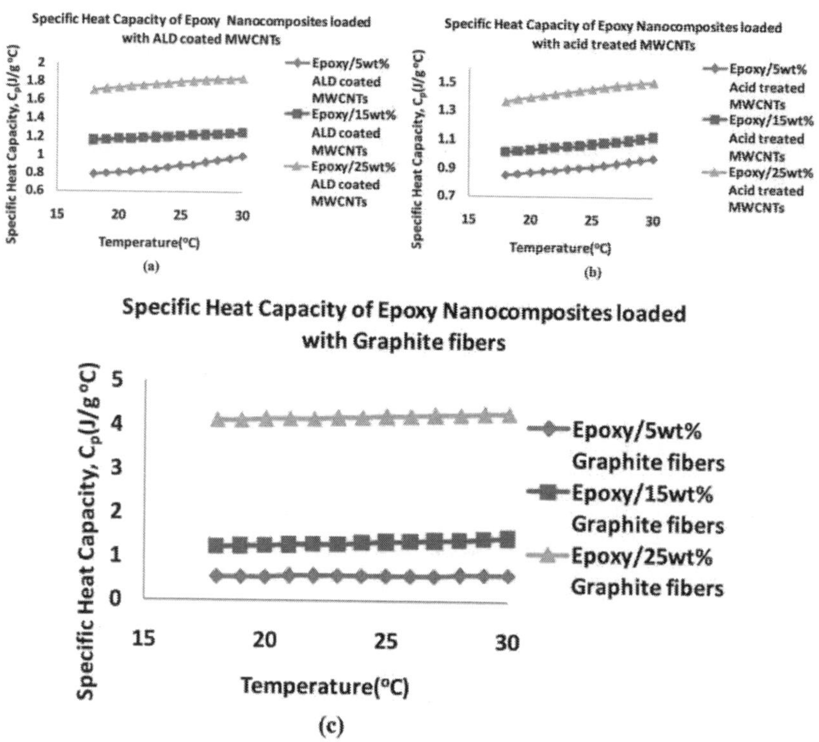

Figure 10. Plots of specific heat capacities of epoxy nanocomposites in the 18°C - 30°C temperature range (5 - 25 wt%) (a) ALD coated MWCNTs (b) Acid treated MWCNTs (c) Graphite fibers.

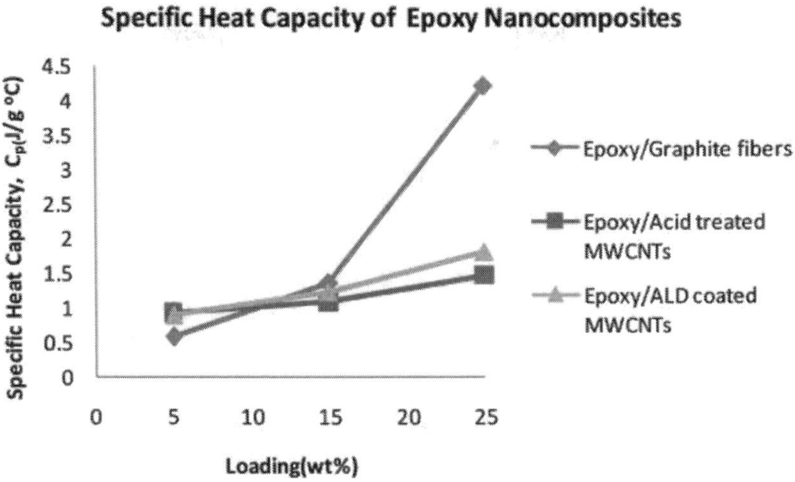

Figure 11. Plots of specific heat capacities of epoxy Nanocomposites at 25°C as the filler loading increases.

Table 5. Thermal conductivity data of epoxy nanocomposites.

Composite	α	C_p	λ
	$[m^2 \cdot s^{-1}]$	$[JKg^{-1} \cdot K^{-1}]$	$[Wm^{-1} \cdot K^{-1}]$
5 wt% CF	2.00×10^{-7}	574.9	1.09
15 wt% CF	2.63×10^{-7}	1339.0	3.41
25 wt% CF	3.80×10^{-7}	4197.0	16.25
5 wt% AL-ALD	1.50×10^{-7}	8850.0	12.30
15 wt% AL-ALD	1.80×10^{-7}	1211.0	2.15
25 wt% AL-ALD	1.77×10^{-7}	1795.0	4.31
5 wt% AC	1.37×10^{-7}	9143.0	11.95
15 wt% AC	1.90×10^{-7}	1070.0	2.14
25 wt% AC	1.40×10^{-7}	1460.0	1.80

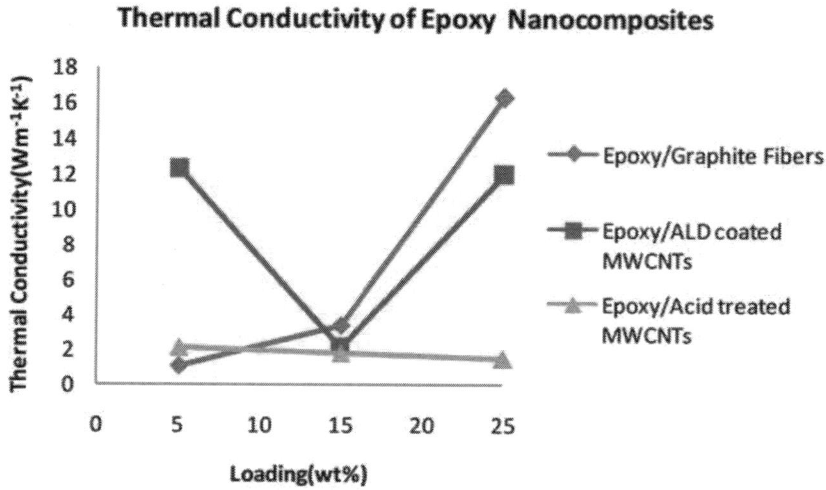

Figure 12. Plots of thermal conductivities of epoxy nanocomposites from 5 to 25 wt% loading.

It can be seen that in general, the thermal diffusivity values are several times higher than those that were obtained in nanocomposites loaded with MWCNTs and graphite fibers. When we look at the thermal diffusivity data that corresponds to the nanocomposites filled with all the three samples of Boron Nitride, we notice that at a loading of 0.1 wt% the uncoated BN filled nanocomposites performed better, followed by the 25 cycle ALD alumina coated BN filled nanocomposite and lastly the 10 cycle ALD alumina coated BN filled nanocomposite with thermal diffusivity measurements of 0.9878 $cm^2 \cdot s^{-1}$, 0.9099 $cm^2 \cdot s^{-1}$ and 0.8643 $cm^2 \cdot s^{-1}$ respectively.

At a loading of 0.3 wt% the 10 cycle ALD alumina coated BN filled nanocomposite reaches a maximum thermal diffusivity of 1.0624 $cm^2 \cdot s^{-1}$ higher than the other two and then drops. At 0.5 wt% loading the 25 cycle ALD alumina coated BN filled nanocomposite reaches the highest thermal diffusivity of 1.1121 $cm^2 \cdot s^{-1}$. These thermal diffusivity values seem to highlight the superiority of Boron Nitride fillers in nanocomposites meant for thermal dissipation.

Figure 14 shows the specific heat capacities of the BN filled nanocomposites in the temperature range 18°C - 25°C. It can be seen that the heat capacities increase with temperature. Figure 15 shows the plots of the specific heat capacities at 25°C for each BN nanocomposite. It can be seen that as the percentage loading increases, the specific heat capacity of the nanocomposite with BN coated with alumina 10 cycles increases more followed by the nanocomposite with BN(uncoated) filled nanocomposite.

These two nanocomposites show an increase in specific heat capacity at 0.1 wt% before it decreases. In the case of the nanocomposite loaded with BN coated with alumina 25 cycles, the specific heat capacity decreases with loading to 0.3 wt% before it rises sharply to 1.419 $Jg^{-1}\cdot{}^{\circ}C^{-1}$ at 0.5 wt%. Table 7 shows the thermal conductivity of the BN filled nanocomposites at each loading. Figure 16 shows the plots of the thermal conductivity data tabulated in Table 7.

Table 6. Diffusivity data of epoxy/Boron Nitride nanocomposites at each loading.

Loading	α (Epoxy/BN)	α (Epoxy/BN10)	α (Epoxy/BN25)
(wt%)	[$cm^2\cdot s^{-1}$]	[$cm^2\cdot s^{-1}$]	[$cm^2\cdot s^{-1}$]
0.1	0.9878	0.86426	0.9099
0.3	0.9163	1.06243	0.9019
0.5	0.9108	0.87340	1.1121

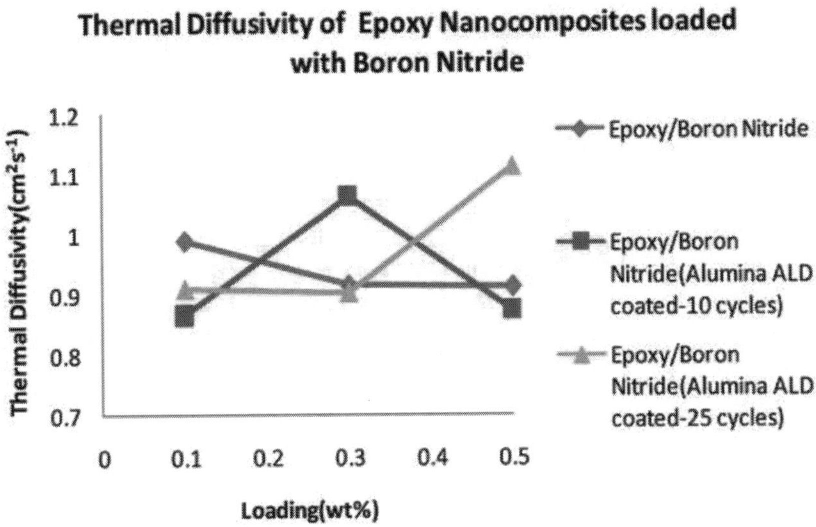

Figure 13. Plots of thermal diffusivity of BN filled epoxy nanocomposites.

The thermal diffusivity, density and specific heat capacity were used to calculate the thermal conductivity of the nanocomposites. The plots indicate significantly high thermal conductivities for all BN nanocomposites as the loading increases irrespective of the number of cycles the BN was ALD coated with alumina. However, the nanocomposites loaded with BN that had been ALD coated with alumina for 25 cycles showed the highest thermal conductivity of 1489.01 $Wm^{-1} \cdot K^{-1}$ at a loading of 0.5 wt%. This high thermal conductivity value suggests that ALD coating of BN with alumina increases their thermal conductivity which ultimately increases the thermal conductivity of the nanocomposites.

There are two main reasons for coating BN particles with an ultrathin film of alumina (Al_2O_3). Firstly, the film results in a particle with minimized reduced thermal conductivity. Secondly, there is significant reduction in the viscosity of the resin in which they are incorporated before curing. This coating therefore increases the surface wettability of the BN particles with the polymer matrix in use and in our case the epoxy resin. As a result, higher loadings can be achieved. ALD (Atomic Layer Deposition) is a coating method which results in films that are conformal, pinhole-free, non-granular and which can be controlled to the level of an atom. ALD splits the involved binary reaction into two self-limiting half-reactions that occur on the surface of the substrate. The ALD deposition of Al_2O_3 in the presence of trimethylaluminum (TMA) and water can be represented by Equation (8).

$$2Al(CH_3)_3 + 3H_2O \rightarrow Al_2O_3 + 6CH_4 \tag{8}$$

$$2AlOH^* + 2Al(CH_3)_3 \rightarrow 2[Al\text{-}O\text{-}Al(CH_3)_2]^* + 2CH_4 \tag{A}$$

$$[Al\text{-}O\text{-}Al(CH_3)_2]^* + 3H_2O \rightarrow Al_2O_3 + 2AlOH^* + 4CH_4 \tag{B}$$

The starred species in equations A and B are surface species representing film growth. Because of the limited number of surface reactive sites, the reaction continues until all the sites have been reacted and no further reaction takes place. This is why the reactions are considered self-limiting or self-terminating. It has been reported that a limit of approximately one monolayer or 1×10^{15} species/cm^2 can be deposited under appropriate conditions. The growth of the Al_2O_3 film is characterized by the recurrence of the reactions (A) and (B) resulting in an ABAB sequence. For every AB reaction cycle, approximately 1 Å of Al_2O_3 film is deposited at 177°C [25-28].

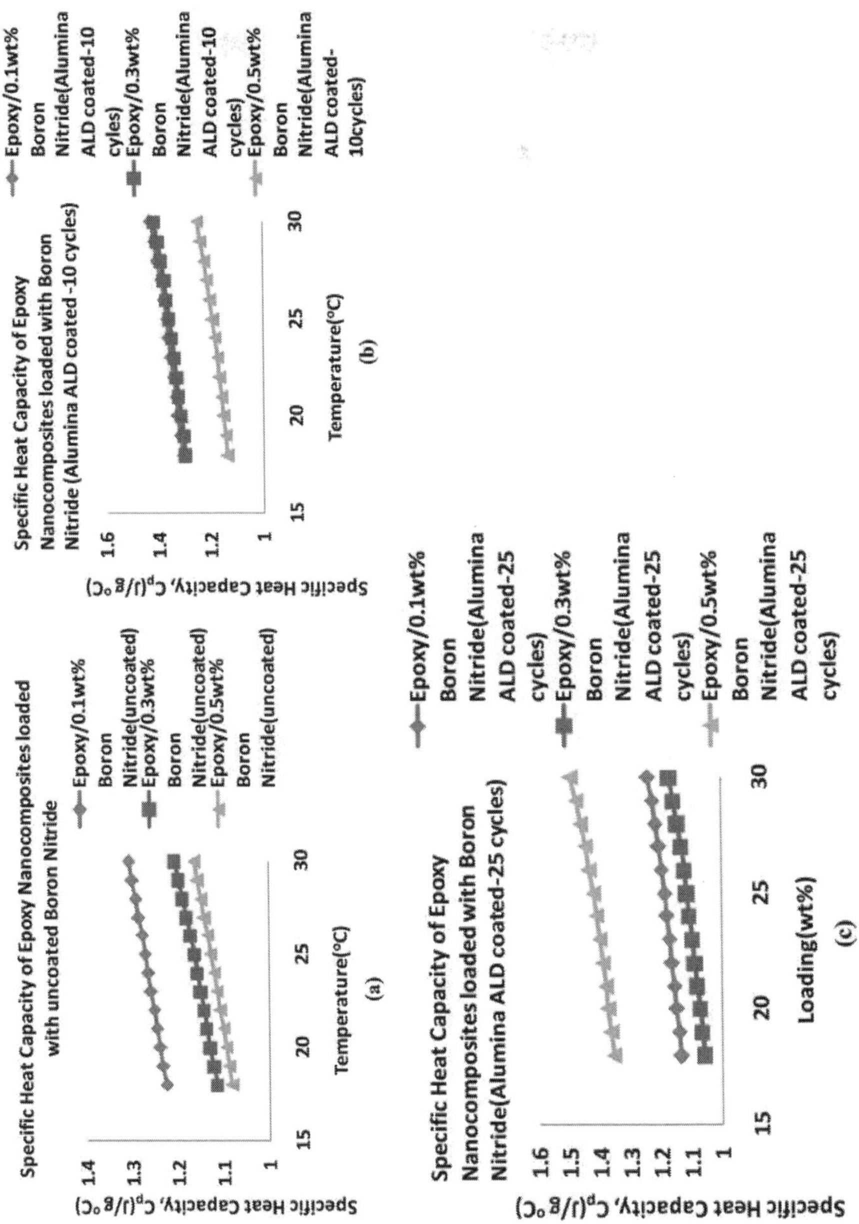

Figure 14. Plots of specific heat capacities of BN filled epoxy nanocomposites in the 18°C - 30°C temperature range (0.1 - 0.5 wt%) (a) BN (Uncoated) (b) BN (ALD coated 10 cycles) (c) BN (ALD coated 25 cycles).

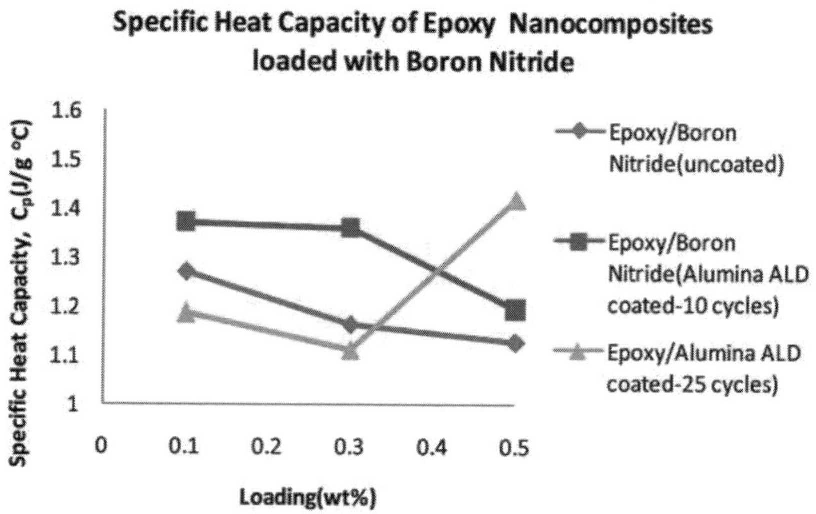

Figure 15. Plots of specific heat capacities of BN epoxy nanocomposites at 25°C as the filler loading increases

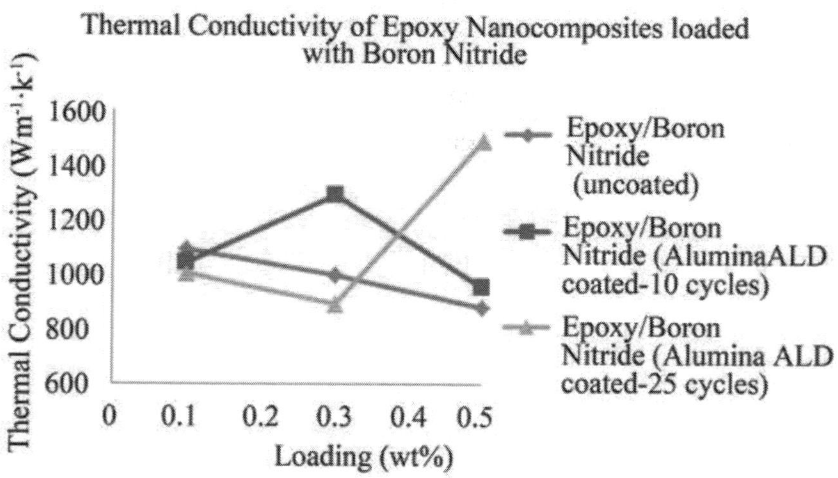

Figure 16. Plots of thermal conductivities of BN loaded epoxy nanocomposites from 0.1 to 0.5 wt% loading.

Table 7. Thermal conductivity data of epoxy nanocomposites.

Composite	α	C_p	λ
	$[m^2 \cdot s^{-1}]$	$[JKg^{-1} \cdot K^{-1}]$	$[Wm^{-1} \cdot K^{-1}]$
0.1 wt%, BN(uncoated)	9.88×10^{-5}	1227.0	1090.77
0.3 wt%, BN(uncoated)	9.16×10^{-5}	1164.0	999.45
0.5 wt%, BN(uncoated)	9.11×10^{-5}	1128.0	881.90
0.1 wt%, BN10	8.64×10^{-5}	1370.0	1039.38
0.3 wt%, BN10	1.06×10^{-4}	1358.0	1291.89
0.5 wt%, BN10	8.73×10^{-5}	1193.0	951.32
0.1 wt%, BN25	9.10×10^{-5}	1186.0	1004.76
0.3 wt%, BN25	9.02×10^{-5}	1112.0	889.21
0.5 wt%, BN25	1.11×10^{-4}	1419.0	1489.01

4. CONCLUSIONS

Nanocomposites were successfully fabricated by blending epoxy with MWCNTs, graphite fibers, and BN. The thermal diffusivities of the nanocomposites were successfully measured by using the Laser Flash method. The nanocomposites loaded with CVD which produced MWCNTs were found to have better thermal diffusivity and thermal conductivity than the other MWCNTs. However, the graphite filled with nanocomposites showed better thermal diffusivity and thermal conductivity than all the nanocomposites loaded with the different kinds of MWCNTs. The thermal diffusivities and thermal conductivities of BN filled with nanocomposites were several times higher than those that were obtained for graphite and MWCNTs filled with nanocomposites. The much better thermal conductivity results that were obtained with BN could be attributed to the ALD Al_2O_3 coating. This coating increases the surface wettability of the BN particles with the epoxy resin.

There is significant reduction in the viscosity of the resin in which they are incorporated before curing due to the wetting of the BN by the resin. A positive consequence of this increased wettability is the possibility of achieving higher loadings. The adhesion between coated particles and the resin is higher than that of uncoated BN.

REFERENCES

1. K. M. Razeeb and E. Dalton, "Nanowire-Polymer Nanocomposites as Thermal Interface Material," In: B. Reddy, Ed., Advances in Nanocomposites-Synthesis, Characterization and Industrial Applications, InTech, Winchester, 2011, pp. 685-706.

2. W. L. Song, P. Wang, L. Cao, A. Anderson, M. J. Meziani, A. J. Farr and Y.-P. Sun, "Polymer/Boron Nitride Nanocomposite Materials for Superior Thermal Transport Performance," Angewandte Chemie International Edition, Vol. 51, No. 26, 2012, pp. 6498-6501.

3. T. Tanaka, M. Kozako and K. Okamoto, "Toward High Thermal Conductivity Nano Micro Epoxy Composites with Sufficient Endurance Voltage," Journal of International Council on Electrical Engineering, Vol. 2, No. 1, 2012, pp. 90-98.

4. S. S. Samal, "Role of Temperature and Carbon Nanotube Reinforcement on Epoxy Based Nanocomposites," Materials Characterization, Vol. 8, No. 1, 2009, pp. 25-36.

5. H. H. So, J. W. Cho and N. G. Sahoo, "Effect of Carbon Nanotubes on Mechanical and Electrical Properties of Polyimide/Carbon Nanotubes Nanocomposites," European Polymer Journal, Vol. 43, No. 9, 2007, pp. 3750-3756.

6. T. Rohr and M. V. Eesbeek, "Polymer Materials in the Space Environment," Proceeding of the 8th Polymers for Advanced Technologies International Symposium, 13-16 September 2005, Budapest, pp. 8-10.

7. W. Gacitua, A. Ballerini and J. Zhang, "Polymer Nanocomposites: Synthetic and Natural Fillers a Review," Maderas. Ciencia y Tecnología, Vol. 7, No. 3, 2005, pp. 159-178.

8. J.-H. Du, J. Bai and H.-M. Cheng, "The Present Status and Key Problems of Carbon Nanotube Based Polymer Composites," eXPRESS Polymer Letters, Vol. 1, No. 5, 2007, pp. 253-273.

9. J. Liu, T. Liu and S. Kumar, "Effect of Solvent Solubility Parameter on SWNT Dispersion in PMMA," Polymer, Vol. 46, No. 10, 2005, pp. 3419-3424.

10. W. E. Jones, J. Chiguma, E. Johnson, A. Pachamuthu and D. Santos, "Electrically and Thermally Conducting Nanocomposites for Electronic Applications," Materials, Vol. 3, No. 2, 2010, pp. 1478-1496.

11. Y. Wang, J. Wu and F. Wei, "A Treatment Method to give Separated Multi-Walled Carbon Nanotubes with High Purity, High Crystallization and a Large Aspect Ratio," Carbon, Vol. 41, No. 15, 2003, pp. 2939-2948.

12. H. Chen, H. Muthuraman and P. Stokes, "Dispersion of Carbon Nanotubes and Polymer Nanocomposite Fabrication Using Trifluoroacetic Acid as a Co-Solvent," Nanotechnology, Vol. 18, No. 41, 2007, p. 415606.

13. M. S. Strano, "Polymer-Wrapped Nanotubes," Nature Materials, Vol. 5, No. 6, 2006, pp. 433-435.

14. Alig, T. Skipa, D. Lellinger, M. Bierdel and H. Meyer, "Dynamic Percolation of Carbon Nanotube Agglomerates in a Polymer Matrix: Comparison of Different Model Approaches," Physica Status Solidi (B), Vol. 245, No. 10, 2008, pp. 2264-2267.

15. H. Fukushima, L. T. Drzal, B. P. Rook and M. J. Rich, "Thermal Conductivity of Exfoliated Graphite Nanocomposites," Journal of Thermal Analysis, Vol. 85, No. 1, 2006, pp. 235-238.

16. W. Nunes, P. Mummery and A. Wallwork, "Thermal Diffusivity of Polymers by the Laser Flash Technique," Polymer Testing, Vol. 24, No. 5, 2005, pp. 628-634.

17. M. Merzlyakov and C. Schick, "Thermal Conductivity from Dynamic Response of DSC," Thermochimica Acta, Vol. 377, No. 1-2, 2001, pp. 183-191.

18. C. Blanco, S. P. Appleyard and B. Rand, "Study of Carbon Fibres and Carbon-Carbon Composites by Scanning Thermal Microscopy," Journal of Microscopy, Vol. 205, Pt. 1, 2002, pp. 21-32.

19. F. Ochanda and W. E. Jones, "Fabrication and Thermal Analysis of Submicron Silver Tubes Prepared from Electrospun Fiber Templates," Langmuir, Vol. 23, No. 2, 2007, pp. 795-801.

20. P. S. Gaal, M. Thermitus and D. E. Stroe, "Thermal Conductivity Measurements Using the Flash Method," Journal of Thermal Analysis and Calorimetry, Vol. 78, No. 1, 2004, pp. 185-189.

21. C. Chiu, J. G. Maveety and Q. A. Tran, "Characterization of Solder Interfaces Using Laser Flash Metrology," Microelectronics Reliability, Vol. 42, No. 1, 2002, pp. 93-100.

22. R. L. Danley, "New Heat Flux DSC Measurement Technique," Thermochimica Acta, Vol. 395, No. 1-2, 2003, pp. 201-208.

23. V. Buryachenko, A. Roy, K. Lafdi, K. Anderson and S. Chellapilla, "Multi-Scale Mechanics of Nanocomposites Including Interface: Experimental and Numerical Investigation," Composites Science and Technology, Vol. 65, No. 15-16, 2005, pp. 2435-2465.

24. J. N. Coleman, U. Khan, W. J. Blau and Y. K. Gun, "Small but Strong: A Review of the Mechanical Properties of Carbon Nanotube-Polymer Composites," Carbon, Vol. 44, No. 9, 2006, pp. 1624-1652.

25. C. Dillon, A. W. Ott, J. D. Way and S. M. George, "Surface Chemistry of Al_2O_3, Deposition Using $Al(CH_3)_3$ and H_2O in a Binary Reaction Sequence," Surface Science, Vol. 322, No. 1-3, 1995, pp. 230-242.

26. W. Ott, J. W. Klaus, J. M. Johnson and S. M. George, "Al_2O_3 Thin Film Growth on Si(100) Using Binary Reaction Sequence Chemistry," Thin Solid Films, Vol. 292, No. 1-2, 1997, pp. 135-144.

27. W. Ott, K. C. McCarley, J. W. Klaus, J. D. Way and S. M. George, "Atomic Layer Controlled Deposition of Al_2O_3 Films Using Binary Reaction Sequence Chemistry," Applied Surface Science, Vol. 107, 1996, pp. 128-136.

28. J. R. Wank, S. M. George and A. W. Weimer, "Nanocoating individual cohesive boron nitride particles in a Fluidized Bed by ALD," Powder Technology, Vol. 142, No. 1, 2004, pp. 59-69.

CHAPTER 7

Geometrical Issues on the Structural Analysis of Transmission Electricity Towers Thanks to Laser Scanning Technology and Finite Element Method

Borja Conde [1,†], Alberto Villarino [2,*], Manuel Cabaleiro [1] and Diego Gonzalez-Aguilera [2,†]

[1] Department of Material Engineering, Applied Mechanics and Construction, School of Industrial Engineering, University of Vigo, 36208 Vigo, Spain

[2] Department of Land and Cartographic Engineering, High Polytechnic School of Avila, University of Salamanca, Hornos Caleros, 50, 05003 Avila, Spain

ABSTRACT

This paper presents a multidisciplinary approach to reverse engineering and structural analysis of electricity transmission tower structures through the combination of laser scanning systems and finite element methodology. The use of laser scanning technology allows the development of both drawings and highly accurate three-dimensional geometric models that reliably reproduce geometric reality of towers structures, detecting imperfections, and particularities of their assembly. Due to this, it is possible to analyze and quantify the effect of these imperfections in their structural behavior, taking into account the actual geometry obtained, different structural models, and load hypotheses proposed. The method has been applied in three different types of metal electricity transmission towers with high voltage lines located in Guadalajara (Spain) in order to analyze its structural viability to accommodate future increased loads with respect that which are currently subjected.

KEYWORDS

non-destructive techniques; structural analysis; Laser Scanning; Finite Element Method; transmission towers; reverse engineering

1. INTRODUCTION

Traditionally, building high-voltage power lines has had few obstacles during their construction phase. Currently, this type of infrastructure is facing a number of setbacks: it has a considerable impact on the environment, on economic activities, and on the expansion of cities, besides its economic cost, including inspections and maintenance. All of these problems have led the companies that use and maintain this infrastructure to consider making the best use possible of the existing lines before placing new lines. Old lines were designed according to the standards of the time in which they were built and they were designed to bear a certain load. In many cases, these towers were designed over forty years ago, so the increased loads that will be placed on them will be far greater than the one for which they were designed. In addition to this fact, the design and execution data of the towers has, in most cases, disappeared, and in other cases, building regulations did not even exist at the time. Due to this, addressing the re-use of existing power lines requires a geometric and structural analysis of the towers to assess their current state.

Formerly, the towers' dimensional analysis was performed through expeditious and manual methods (through the use of a gauge and a measuring tape) that required direct contact with the structure and, therefore, meant high risks and high costs. Afterwards, in search of a remote non-invasive measuring method, classic topographic measuring allowed thorough, notably intense, field work taking angular measurements and determining singular points indirectly through angular intersections. More recently, the existence of reflectorless electromagnetic surveying equipment has allowed direct measurement of distances and angles from a single point, making field work easier and more efficient, although it solely focuses on extracting unique and specific measurements determined by the topographer [1]. This has meant great uncertainty upon the elements of the tower, since the data was only taken at the point where the measurement is performed. In order to have the full representation of the geometry of the structure, in the last years laser scanning has presented as an interesting solution [2,3,4,5], due to the fact that they generate dense real-time point-clouds of the tower's geometry from a distance [6]. However, one of the major limitations of these terrestrial geotechnologies is the overall height of the tower, impossible to cover completely from the ground, which has led to the use of robotic unmanned aerial systems that take aerial images and, through photogrammetric procedures, obtain dense three-dimensional models of this type of infrastructure [7].

As for structural assessment, this kind of structure has been analyzed from different points of view as presented in the literature: the effects of loading in the stability of the tower [8,9,10,11]; the effects of the stiffness of connections [12,13,14]; and causes of failure [15,16,17]. No previous work was performed in order to evaluate the effect of geometric imperfections such as misalignments of structural members at joints or assembly imperfections.

Therefore, in view of what is mentioned above, this paper presents a non-destructive multidisciplinary approach that is articulated in two stages and that analyzes geometrically and structurally electricity transmission tower structures:

1) The first stage will address a detailed geometric description of the structure (reverse engineering) using a terrestrial laser scanner system, performing an "as built" model that provides information on the structure's most relevant data, such as geometry, type, and dimensions of the metallic profiles and their assembly drawings.

2) Secondly, and taking the geometric model obtained by the laser system as a starting point, three different structural finite element models will be developed: one model will have an ideal geometry considering the nodes of the transmission tower to be pinned, supposing that this model was the one that was used in the original design of the towers. A second model will be developed with the ideal geometry of the previous model, but considering certain continuous elements of the transmission tower to be rigidly connected at both ends. The third model will use the real geometry obtained from the laser scanner (taking into account all possible initial imperfections as, for example, misaligned structural members at joints) and will also consider the stiffness behavior of the continuous elements of the tower. For its analysis the current Spanish standard for the design of such towers will be taken into account [18]. The analysis will be carried out in a linear elastic regime with the software SAP2000 [19] to obtain data of the displacements and stresses

in transmission tower members for each load case established by the current codes. It is expected to obtain conclusions about the performance of these different structural approaches and, therefore, conclude which is the most appropriate modeling strategy in structural assessment procedures when an increase in loads to accommodate new services is demanded.

The paper is outlined as follows: after the introduction, in Section 2, the characteristics of the towers' considered structural models are described. Section 3 details: the equipment used, the production of the geometric models through computer aided design (CAD) software, and the data used in structural analysis. In Section 4 the results are analyzed and end with the reached conclusions.

2. STRUCTURAL MODELING OF TRUSS STRUCTURES

A transmission tower could be considered as a three dimensional truss structure [20], which is comprised of a reticular structure formed by discrete elements (bar or rods), joined together at their ends by means of connections without friction and designed to withstand the external forces by means of axial internal forces in their members. The idealized structural model used for the study of this kind of structure is usually based on the following assumptions [21,22]:

1) Individual elements or rods are joined together at their ends by means of connections with no friction, which means the nodes transmit forces but do not transmit moments.

2) The centroidal axis of each member is straight and matches the line that joins each end of the member. The axis of all of the members that end in the same node is cut at a single point; otherwise moments will appear in these members so that the node could be at static equilibrium.

3) Whenever external loads are applied in the nodes, all the elements that comprise the structure are subjected to tensile and compressive forces, since there is no friction at the connections. This means that the self-weight of the elements is replaced with loads applied at their ends. According to [23] bending caused by direct wind loads on the structural elements can be omitted.

4) The cross section of each member has a negligible area compared to its length.

5) The self-weight of the elements is negligible, since the loads supported by the structural elements are usually large in comparison to their weights.

Under the fulfillment of these assumptions, the structural elements are exclusively subjected to axial forces. For the case studies later reported in this paper, given the age of the towers and the virtual inexistence of structural calculation software at that time, it is logical to assume that they were either calculated manually, using the so-called Ritter method [21,22] or graphically, using the Cremona method [21,22].

However when looking at the analyzed tower's "as built" model we found out that assumption 1 and 2 are not true in the case studies herein presented:

1) In the case of members of the towers, their connections are bolted and thus they are not actually frictionless nodes. Additionally, some members, such as the truss chords, are continuous elements, which can evidently transmit moments from one side of a node to another [24].

   Thus, these nodes are assumed to be rigidly connected behaving as elastic embedment (or simply rigid joints), which cause local bending moments due to rotations that take place as a consequence of the global deformation of the structure. Such effect causes the so-called secondary bending moments and consequently a secondary stress state, which are juxtaposed to the primary stresses derived from axial forces.

2) Furthermore, truss members that end in the same node do not always cut it at the same point, leading again to an apparition of secondary bending moments. Such bending moments depend upon both proportion of the misaligned or eccentricity and stiffness of the elements.

Both circumstances are observed in all towers herein analyzed. Figure 1 corresponds to an example of the aforementioned issues upon one of the case studies later described (Tower 1). Furthermore, as it can be appreciated, the ideal situation of perfectly-aligned elements assumed in the initial stage of structure design and calculation can suffer variations during its construction on site. This is shown in the "as built" model, obtained with the laser scanner, where small deviations in the horizontality of several truss members and even a small deviation in the vertical alignment of the main body of the tower can be appreciated (Figure 1).

All of these issues question the use of a model with ideal pinned connections, which was possibly the used method at the time of the structure's design and calculation, and whether it is a valid model to perform further calculations at the present time. Therefore, in view of the constraints and considerations discussed above, three different structural models of the towers are tested in order to compare the results with each other and draw conclusions about the deformations and the actual stress state of the tower's elements, all under different load hypotheses established by current standards [18].

1) Model with pinned joints and ideal geometry (Model 1). This model is supposed to be the one that was used for the design of the towers. A three-dimensional model with all members pin-jointed at nodes was developed that took into account all the previously-mentioned assumptions. Accordingly, the adoption of certain assumptions about the geometry provided by the laser scanner was needed in order to adapt the model to an ideal geometry. More specifically, it was assumed that all members concurring in one node were overlapping in a single common point since, in reality, these members were assembled more erratically. For this, the criterion used was to assume that the main vertical members of the tower and the horizontal members were fixed in their position, and the corrections changed the length and exact location of the diagonal braces that connect at each joint. Additionally, the geometry was simplified in certain areas of the horizontal sections of the towers, ignoring the bend and the inclination. Accordingly, vertical deviation of the main bodies of the towers was not taken into account. Finally, the obtained "as built" model measurements of certain modules that form the main body of the towers are simplified to theoretical values. For example, when certain lengths result in 1.99 m or 2.01 m, the value is simplified and considered to be 2 m. For clarity, Figure 2 presents a scheme with the principal structural parts of a transmission tower.

2) Model with rigid joints and ideal geometry (Model 2). According to the data proportioned from the "as built" model, assumption 1 is not valid because certain elements, such as truss chords, are continuous elements. Therefore, a second model was elaborated which considers an ideal geometry but where truss chords and horizontal members of the towers are considered rigidly connected at both ends and all other structural components (truss elements) are treated as pin jointed. In this way, the originated secondary bending moments and, consequently, secondary stresses can be taken into account and their impact in comparison to modeling the tower with all ideal pinned joints quantified.

3) Model with rigid joints and real geometry (Model 3). In contrast with the two previous models, this approach considers all of the real geometric singularities of the structure and includes them into the simulations; for example, it includes horizontal deviations of truss

chords, the vertical deviation of the tower's main body, as well as assumptions 1–3 obsolescence. To create such a structural model, first a 3D wire-model was produced using the "as-built" model as a dimensional foundation. Secondly, it was imported as a "dxf" file into the structure calculation software SAP2000 [19].

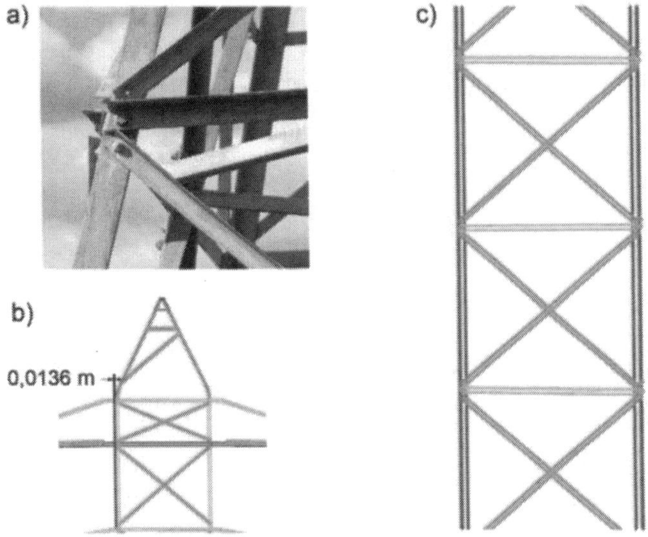

Figure 1. (**a**) Detail of the bolted connections with the inclusion of the eccentricity present at the linkage between profiles; (**b**) vertical deviation of the main body of tower; (**c**) horizontal deviation in the members of the main body of tower.

3. METHODOLOGY

3.1. Geometric Digitalization

A laser scanning survey was conducted in order to generate a 3D model accurately describing the structure of the transmission towers. A time-of-flight (ToF) terrestrial laser scanner (TLS), Leica ScanStation2, was used for recording external geometry. This scanner covers a field of view of 360° in the horizontal direction and 270° in the vertical direction, enabling the collection of full panoramic views. The distance measurement is obtained with a nominal accuracy of 4 mm at 25 m range. The horizontal and vertical angular accuracy is of 60 μrad (3.8 mgon). The diameter of the laser spot is 4 mm at 50 m. According to the technical specification of the instrument, it has a maximum acquisition rate of 50,000 points per second. The scanner incorporates a dual axis compensator, so the vertical Z-direction is perfectly defined during data acquisition. Due to the complexity of transmission towers and their heights, four scanner stations with a resolution of 5 mm at 25 m were required to enclose the whole structure. The resulting 3D point cloud (about 6 million points per tower) contains geometric data, normally given in Cartesian coordinates (XYZ), as well as the intensity values (I). This intensity measurement is referred to as the amount of reflected radiation with respect to the emitted radiation. Typically, this value is normalized in the range of 0–1 or 0–255, 8-bit format in our case. According to the principles of interaction between electromagnetic radiation and matter, the intensity values directly depend on the physical characteristics of the object surface, the wavelength of the incident radiation, and the distance between the laser and the object.

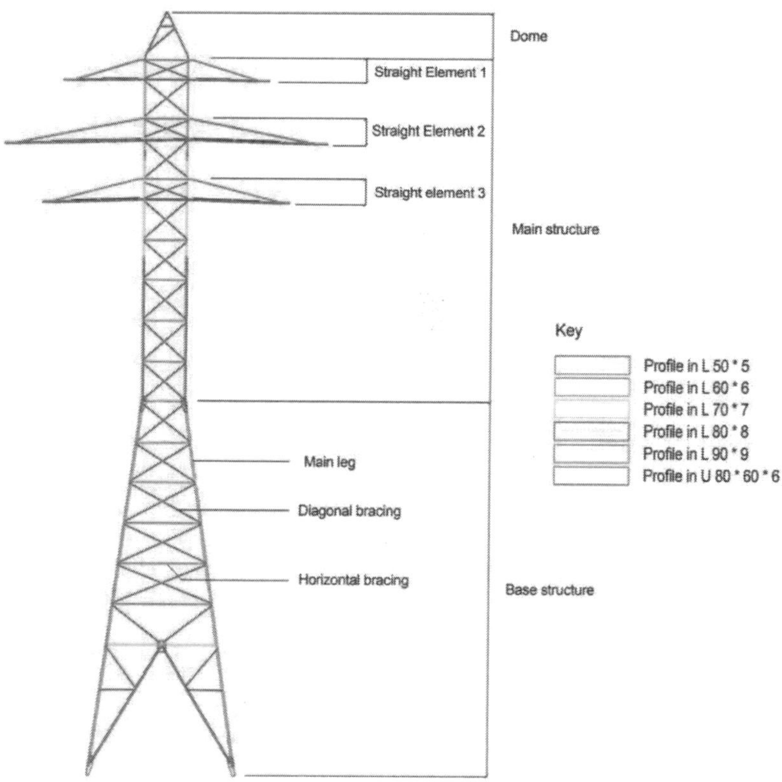

Figure 2. Scheme of main structural parts of transmission tower upon draws of one of the cases studies herein analyzed (Tower 1). Data was obtained from laser scanning technology according to the methodology presented in Section 3.

3.2. Geometric Modeling

The geometric modeling of the transmission towers was performed following four steps:

1) Cleaning and segmentation of point clouds in order to remove undesired data, such as reflections, noise or sensor artifacts. This step was performed manually.

2) Alignment of the point clouds from each scan under a common coordinate system. An automated registration method, iterative closest point (ICP) [25], was applied, supported by the identification of matching points and minimizing the Euclidean distance between corresponding point clouds. Initial approximations (three points) were manually identified by the user, trying to guarantee a good distribution around the area of interest and along the three main directions (X, Y, Z). A solid-rigid transformation based on the three points identified was executed. Afterwards, an automatic iterative process to align the different scans was performed taking the Euclidean distance as a minimization criterion.

3) Generating cross-sections and technical drawings of the electrical towers, focusing on the steel profiles that make up each section of the tower and the arrangement of the connections used to define the linkage of these profiles. Different profiles were automatically generated along each main direction (X, Y, Z) in order to obtain vector information of the main sections of the towers and, thus, initial approximations to support the technical drawings and CAD model generation.

4) Obtaining a computer aided design (CAD) model. Since the structural analysis based on a FEM model does not cope with dense laser models, an important step which allows us to pass from the 3D point clouds to a solid geometric model was performed. This step consists in extruding the sections obtained in the step before along its normal direction. Manual interaction is required in this step in order to solve the different intersections between profiles and their connections. In addition, specific existing libraries based on standard steel profiles (i.e. L-shaped and U channel profiles) were used for modeling the towers. Geometric modeling was done using Geomagic Spark, 2013 version.

3.3. Structural Analysis

The three towers are formed by angular steel profiles of different dimensions, and given the age of the towers and only for the purpose of the methodology developed in this article, we assume the lower specification for a steel material enabled by [26], type S-235.

This brings the following mechanical properties: Young's modulus of 2.1×10^8 kN/m^2, specific weight of 76.9729 kN/m^3, Poisson's coefficient of 0.3, and yield stress of 235 MPa.

Furthermore, the data of the power line, support type of the tower, and the mechanical characteristics of the electrical drivers are detailed below:

- High voltage power line with rated voltage of 132 kV and 50 Hz AC
- Two duplex-circuit line with alignment support.
- Span: 300 m between supports.
- Electrical driver of aluminum galvanized steel, type LA-280.

The boundary conditions of the three towers are assumed to be articulated supports in each of the legs that make up the outer frame of the towers, so that they have only limited movements according to the global axes (X, Y, Z). The constraints upon the members are explained in each of the structural models discussed above.

The different load conditions are obtained according to [18]. The following descriptions summarize the loads, always bearing in mind that the towers are located in the province of Guadalajara (Spain) and that they are power lines with alignment support and with suspension insulator strings [18].

- Permanent loads: The self-weight of the steel profiles that comprise the towers, electrical conductors, fittings, insulators, and the grounding wire.
- Wind load: It acts upon steel members of the towers, the insulators, and the suspension insulator strings.
- Imbalance of tensile forces: A longitudinal force equivalent to 25% of all unilateral tractions of electrical drivers and grounding wire. This tensile force will be applied at the point where the electrical conductors and the grounding wire are attached to the support, thus taking into account the torsion that these forces could create.

- -Electrical conductor failure: A unilateral tensile force related to a single electrical conductor or a grounding cable's failure. The minimum admissible value for the failure is 50% of the broken cable's tension in the power lines that have two conductors per phase.

Taking into account aforementioned load patterns, the current standards [18] refer to certain calculation hypotheses that establish the load cases, shown in Table 1.

4. EXPERIMENTAL RESULTS AND DISCUSSION

4.1. Case Studies

Since this study arises as a consequence of the need to analyze the structural viability of a series of electricity transmission towers that serve as support to an old power electricity line located between the cities of Guadalajara and Torija (Spain), three different cases studies were chosen for the development of the aforementioned methodology.

The electricity transmission towers chosen correspond to a type of tower known as "support alignment" which are disposed over different sections of electricity line. Additionally, in order to extend the study over different structural typologies, each tower corresponds to a different morphology.

The first tower is formed by both a main body support and another principal body (comprising horizontal bracings and diagonal bracings according to a St Andrew's disposition) and three horizontal symmetrical bodies for the support of the cables. The second tower only has a support body (formed by horizontal bracings and secondary diagonal bracings according to a St Andrew's disposition) and three asymmetric horizontal bodies. The third tower is similar to the second one, with exception in the diagonals forming the support body which are not arranged according to a St Andrew's disposition.

In order to clarify such load cases considered, in Figure 3 is detailed upon finite element model of one of the case studies herein analyzed (Tower 1), the arrangement of the loads in each of the three scenarios previously commented. Similarly, loads are arranged in Towers 2 and 3.

Figure 4 shows a photograph of the three towers that composes cases studies described above.

Table 1. Load cases considered in structural analysis of towers.

Tower Type	Force Direction	Hypothesis 1	Hypothesis 2	Hypothesis 3
Alignment support and suspension insulator strings	Vertical	Permanent loads, considering the electrical conductors and the grounding cables to withstand wind load according to a 120 km/h wind speed		
	Transversal	Wind load (120 km/h) on electrical conductors, cables, grounding cables and supports of tower	Not applicable	Not applicable
	Longitudinal	Not applicable	Imbalance of tensile stress	Electrical conductor and grounding cable failure

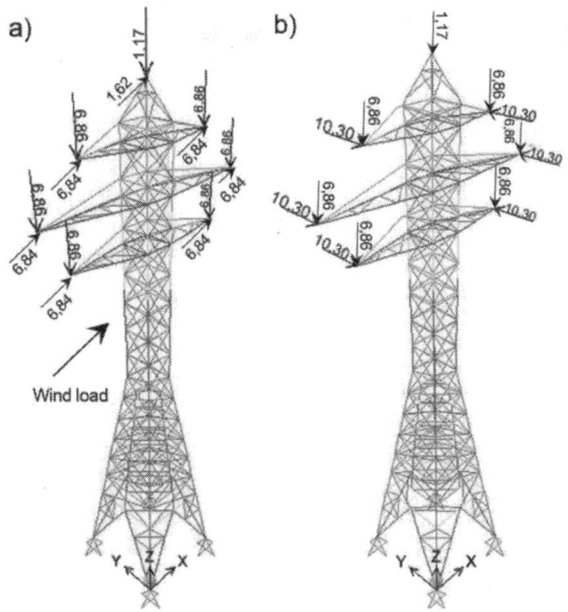

Figure 3. Loads cases considered in structural analysis. (**a**) Hypothesis 1: wind; (**b**) Hypothesis 2: Imbalance tractions; (**c**) Hypothesis 3: electrical conductor failure.

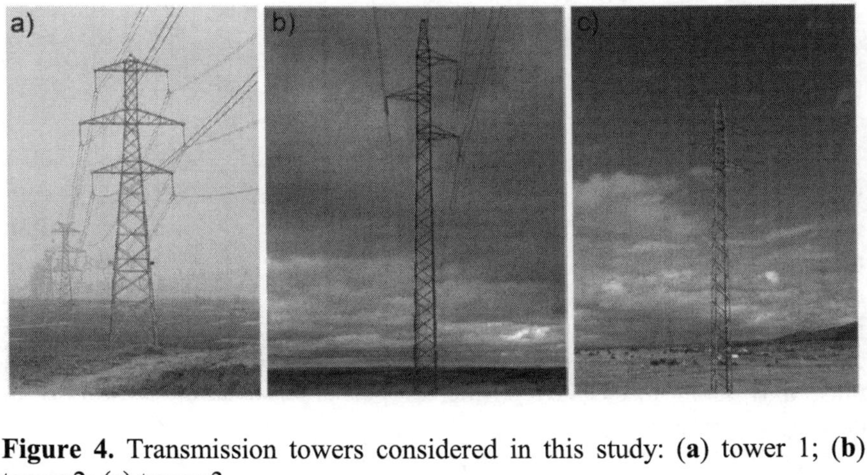

Figure 4. Transmission towers considered in this study: (**a**) tower 1; (**b**) tower 2; (**c**) tower 3.

4.2. Geometric Modeling

Following all steps described in Section 3.2, the point cloud data obtained as a result of a laser scanning survey was subsequently transformed into a CAD model valid for its implementation in the finite element software package SAP2000.

This is a key step required in this kind of reverse engineering process, since data obtained from laser scanning technology do not represent any valid information by itself for the purpose of finite element analysis without suitable data processing [27].

Therefore, taking this into account, CAD models for each one of towers analyzed together with drawings about its current disposition and assembly information were obtained.

Figure 5 shows CAD wire models obtained for each one of towers analyzed. Once such geometrical models were obtained, they were directly imported as a DXF file to SAP2000 software for the finite element analysis stage.

Main geometric data concerns to dimensions of the base, height of the tower, and length of horizontal bodies are displayed in Table 2. The length of the horizontal bodies of the towers is measured from the main body of the tower up to the farthest node. Tower 1 has three horizontal bodies with different dimensions, while in towers 2 and 3, all the horizontal bodies have similar dimensions.

4.3. Structural Analysis

As was previously indicated, finite element models of transmission towers were analyzed in SAP2000 software. Within this package, frame elements were chosen so that stiffness against rotations could be considered in all of those nodes assumed to behave as rigid joints. For all other cases where moments will not be considered (as, for example, in diagonal members pin jointed to truss chords), releases end options in nodal degrees of freedom could be imposed for transforming frame elements to truss elements and, thus, only axial forces be considered. For all structural models herein developed, analysis was carried out considering linear elastic behavior.

Table 3 shows the number of frame elements and the number of degrees of freedom for each of the finite element model developed, taking into account the type of transmission tower structure and structural model approach.

Table 2. Main geometric data for transmission towers analyzed.

Tower	Base Dimensions (m)	Height (m)	Length of Horizontal Body (m)
			3.95
1	7.25 × 7.25	37.25	6.25
			5.0
2	1.5 × 1.5	18.50	2.0
3	1.5 × 1.5	18.50	2.15

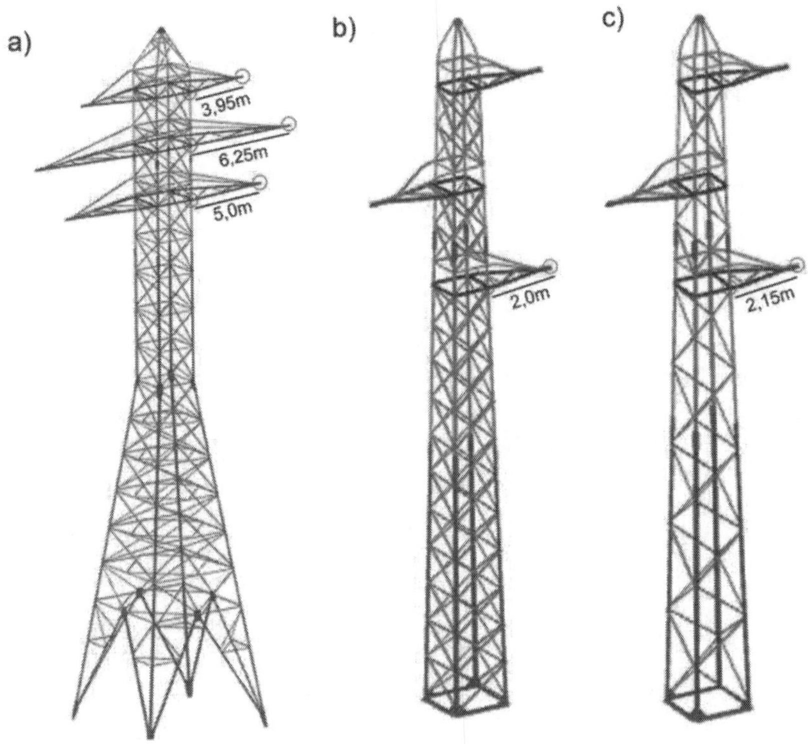

Figure 5. Geometrical CAD wireframe models of transmission towers analyzed: (**a**) tower 1; (**b**) tower 2; and (**c**) tower 3.

4.3.1. Displacements

In this section, displacements [12,14,28] experimented by each tower under different load cases and structural models are analyzed and discussed.

To carry out the analysis and comparison, representative points of the three towers associated with the nodes of the horizontal bodies and the upper node of the dome were selected. Figure 6 shows an example of these nodes together with the displacements experimented for each tower under Loading Case 1 (wind load) and structural model considering real geometry for upper node of the dome (Node 1).

Table 3. Number of frames elements and degrees of freedom for each one of the three different structural models considered upon cases studies analyzed.

Tower	Model 1		Model 2		Model 3	
-	Frame Elements	Degrees of Freedom	Frame Elements	Degrees of Freedom	Frame Elements	Degrees of Freedom
1	419	3320	419	3896	683	6772
2	241	1909	241	2348	321	3308
3	181	1448	181	2100	226	2408

Table 4 reports both the values of maximum displacements obtained for Model 3 (real geometry) and the node where they take place (one of the aforementioned) for the different loading scenarios considered in each one of the transmission towers analyzed. Table 5 represents a comparison (in absolute value) between maximum displacements for Models 1 and 2 with respect to Model 3 (considered as the most accurate).

In both tables, the structural model considered is represented by “M” followed by a number indicating the corresponding model described in Section 2. Load cases considered are indicated by “H” denoting each one of the hypothesis described in Section 3.3 and Table 1 is applicable.

The results displayed in Table 5 show that principal differences always take place in all towers under Hypothesis 1 (wind load case). The differences found between Model 3 and Model 1, reaching 95 mm in the global X direction is remarkable.

There are also significant differences for all towers, although less than aforementioned for displacements of nodes in the global Y direction under Hypotheses 2 and 3 (Imbalance tractions and cable break). The reasons that can explain these differences may be due to the following factors:

- Consideration of the initial small vertical deviations of the towers’ main bodies along with the application of wind forces in Hypothesis 1, precisely in that direction, may have accentuated the displacements in global X direction obtained in Model 3.

- The fact of considering the true connections of the profiles onto the nodes (misalignment) has a drastic consequence upon the behavior of the structure since it directly affects to its stiffness.

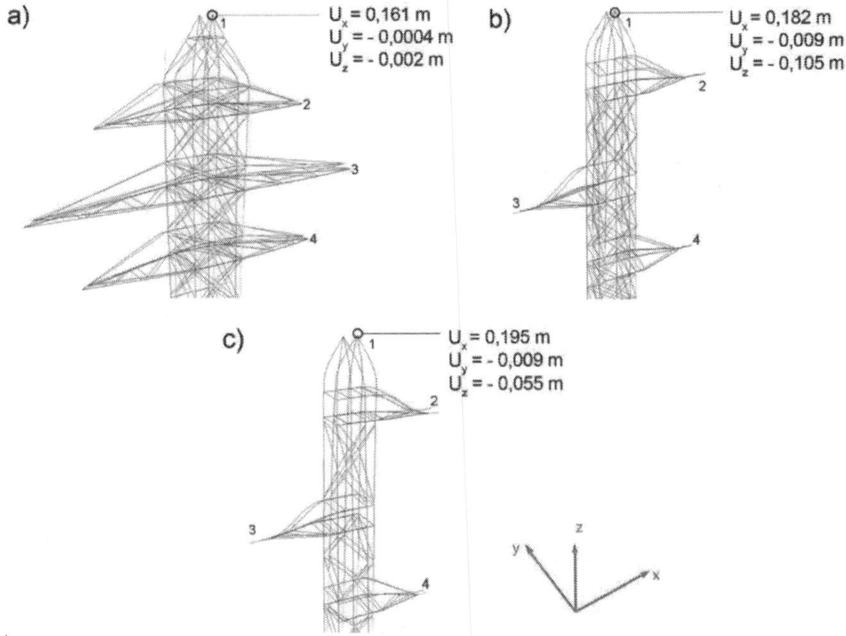

Figure 6. Nodes considered for analysis of displacements according to the global axis directions in each tower and structural model analyzed. (**a**) tower 1; (**b**) tower 2; (**c**)tower 3.

Table 4. Values of the maximum displacements experimented for Model 3 in all transmission towers and load cases considered with the indication of the node where they take place.

Tower	Model	Displacement Ux (mm)/Node			Displacement Uy (mm)/Node			Displacement Uz (mm)/Node		
		H1	H2	H3	H1	H2	H3	H1	H2	H3
1	M3	161/1	14/3	12/4	8/3	162/3	140/3	132/3	8/2	8/2
2	M3	182/1	44/2	11/2	9/1	88/1	125/3	123/2	11/2	15/2
3	M3	195/1	52/2	15/2	9/1	97/1	146/3	85/2	21/3	18/2

Table 5. Comparison between displacements experimented by Models 1 and 2 with respect to Model 3 in all transmission towers and load cases considered.

Tower	Models Compared	Maximum Deviation Ux (mm)/Node			Maximum Deviation Uy (mm)/Node			Maximum Deviation Uz (mm)/Node		
		H1	H2	H3	H1	H2	H3	H1	H2	H3
1	M3 front M1	93/1	6.3/3	4/4	5/3	52/3	62/3	27/3	4/2	4.6/2
	M3 front M2	95/1	6.3/3	4/4	5/3	65/3	78/3	29/3	4/2	5/2
2	M3 front M1	66/1	8.3/2	7/2	4/1	44/1	39/3	35/2	4/2	9/2
	M3 front M2	80/1	9/1	8/2	6/1	43/2	48/3	38/2	5/2	10/2
3	M3 front M1	75/1	9/2	6/2	7/1	38/1	37/3	38/2	3/3	5/2
	M3 front M2	65/3	7/1	7/2	4/2	32/2	42/3	43/3	5/3	6/3

In Models 1 and 2 "ideal nodes" connecting various elements mobilize stiffness of all concurrent elements (truss chords, diagonal, and horizontal members): axial stiffness in Model 1, and axial and bending stiffness in Model 2; in Model 3, however, it does not occur equally.

When considering real geometry, the existence of "intermediate" nodes inserted into the truss chords cause local bending moments and, thereby, additional rotations that accentuate local deformations of the structures. Therefore, we can state that the improper execution of the connections leads to a less stiff structure and may be the main cause of the differences found between the displacements obtained in the three towers for global X direction under Hypothesis 1 and for global Y direction under Hypotheses 2 and 3.

A representative example of this behavior can be seen in Figure 7, which shows the deformed shape of tower 1 for Models 2 and 3 under Load Case 3 (electrical conductor break), which subjects the body of the tower to a torque.

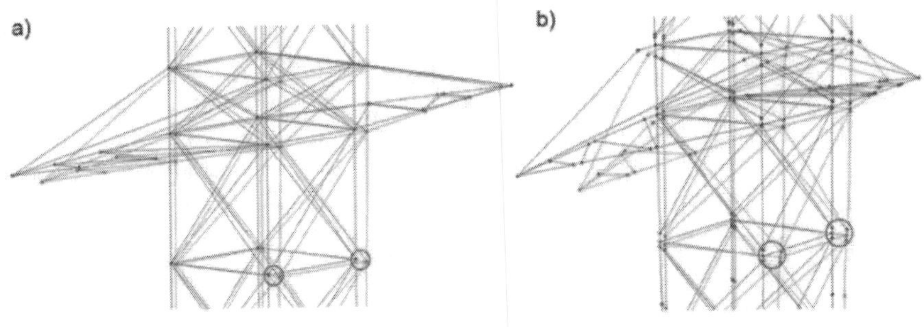

Figure 7. Detail of the different structural behavior of the Tower 1 under Load Case 3 in Model 2 (**a**) and Model 3 (**b**). Undeformed shape is shown in blue and deformade shape in red.

4.3.2. Stresses in Structural Elements

Finally, a comparison regarding the stresses in structural elements [12,14,15,16,29] is also established. Table 6 shows, based on the finite element results obtained, the maximum normal stresses for each transmission tower and its respective structural model. Particularly, it details for each tower, and for each model, the maximum normal stress in which member it occurs, and under which loading case is developed.

Table 6. Structural elements with maximum normal stresses for all transmission towers analyzed and structural models considered.

Tower	Model	Maximum Stress (Mpa)/Frame Element		
		Hypothesis 1	**Hypothesis 2**	**Hypothesis 3**
	1	215.15/90	165.26/401	195.56/468
1	2	225.36/90	170.23/401	205.21/468
	3	252.30/696	185.96/433	230.20/696
	1	218.96/380	155.50/442	221.32/405
2	2	222.56/380	163.89/442	232.45/405
	3	234.21/178	181.78/439	245.63/189
	1	222.25/272	167.25/353	221.45/107
3	2	232.63/272	175.63/353	226.12/107
	3	252.58/272	195.56/353	233.16/104

Based on Table 6, it can be observed that the differences in maximum normal stresses between Model 1 and Model 2 for all towers are quite small and the maximum of these values are always produced in the same members for both models. On the contrary, significant differences appear when compared to Model 3. Consider the case of Tower 1 and under Load Case 1 where the differences is approximately 37 MPa.

Moreover, it could be observed that the maximum stresses in Model 3 no longer occur at the same members that for Models 1 and 2. This is due to the singular arrangement of nodes in Model 3 (members do not intersect at a single point) which leads to a different discretization of frame elements compared to Models 1 and 2.

Rising stresses in Model 3 are mainly due to secondary stresses caused by the additional bending moments derived from small geometric eccentricities at diagonal and horizontal member's connections upon truss chords which are neglected in Model 1 and Model 2.

The higher the eccentricity is at the connections, the greater the induced bending moments will be. Likewise, the level of induced secondary stresses will be influenced by the stiffness of the truss chords; the greater is their stiffness the greater bending moments will be induced.

This is perhaps the cause of the observed differences for Tower 1 under Load Case 1, where the stiffness of the truss chords relative to the overall stiffness of the whole structure is greater than in the other two towers, thus providing greater increased stresses (37 Mpa) with respect to the theoretical Model 1.

It should be also noted that for Model 3, under certain load cases in all towers, some elements exceed the yield point of the material. This circumstance is highlighted in Figure 8, where for tower 1 and under Load Case 1, the element with maximum normal stress reaches 252 MPa, exceeding the elastic limit of the material in 17 MPa. Figure 8 also shows a representative image of the above discussed; the effects of improper execution of nodes upon truss chords and, consequently, the different discretization of frame elements in the model with ideal geometry and those based on real geometry.

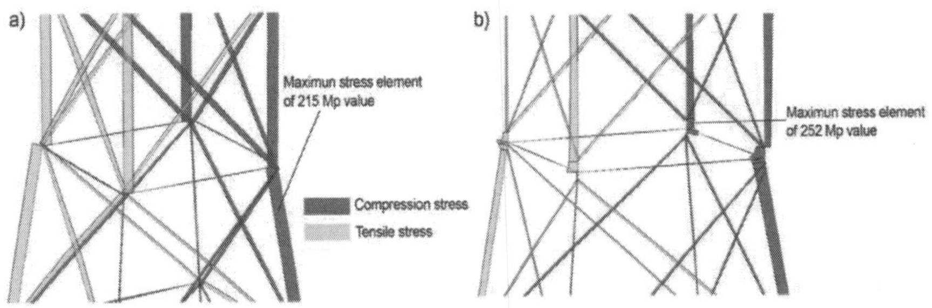

Figure 8. Normal stresses distribution onto Tower 1 for different structural models. (**a**) Model 1, with ideal representation of the nodes and maximum stress below the yield limit of steel; (**b**) Model 3, including the improper executions of the nodes and maximum stress exceeding the yield limit of steel.

As for results involving values of maximum normal stresses, they should be analyzed carefully. Due to the age of the structures, assumed material properties have been chosen according to the minimum value specified by the current regulatory codes; however, these values do not have to match the actual properties of the structure. Accordingly, appropriate experimental tests should be carried out in order to improve their characterization and, thus, derive proper conclusions about the real current safety state of the structure. Note, also, that the structural analysis of transmission towers was carried out assuming a linear elastic behavior.

Nonetheless, the results herein obtained could be considered as acceptable, bearing in mind that the present work is focused in defining an overall methodology able to detect the geometric imperfections present in electricity transmission tower structures by means of precise laser scanning systems and the procedures to incorporate them in structural models based on finite element methods.

5. CONCLUSIONS

Terrestrial laser scanning enabled performing a non-invasive remote survey of different transmission tower structures; it should be noted that these are objects of great complexity, not only for their size, but also by their geometry and their high heights. This technology allowed detecting significant imperfections in terms of connections between the members at the nodes, loss of verticality of the towers and the lack of horizontality in the horizontal bracings. The aforementioned imperfections motivated the consideration of different structural models for the towers, in order to analyze how this affects their structural behavior. To that purpose, three different models have been carried out: first, a model with an ideal geometry and considering perfectly pin jointed nodes (supposed model in original calculations of the structures); second, a model with ideal geometry, but taking into account real existing continuity in some profiles by means of rigid connections and pin jointed connections of all others elements upon them; and finally, a third model similar to that previous, but with a real geometry incorporating all imperfections obtained from laser scanning data.

When analyzing the results obtained in terms of displacements and stresses yielded by the different structural models considered, significant differences were observed. At those nodes considered for the comparison of displacements, differences between models with ideal geometry, and models with real geometry reached several centimeters, becoming the highest value (for Node 1 in Tower 1 under wind loading) 9.5 cm respect to the X global direction.

The study of stresses brings some other conclusions. Considering models with ideal geometry and considering pinned joints (Model 1) or rigid joints (Model 2) no significant differences were found. Indeed, maximum normal stresses in elements always take place in the same elements for both models. On the contrary, the model which accounts for real geometry (Model 3) presented notable differences when compared with their respective idealized models. Notorious increases in stresses were detected under certain loading conditions, even reaching the elastic limit of the steel in some occasions.

Differences observed in displacements, stresses in elements and, thereby, whole structural behavior of towers suggest that a detailed survey and conscientious structural analysis has to be carried out when these type of structures will be required for future uses as, for example, new communication services that increase their service loads.

Further studies could contemplate performing nonlinear analysis to extend and improve the results herein obtained, either by considering geometric nonlinearity effects such as P-Delta effects and plastic behavior of steel material. Moreover, due to the nature of the structures (quite slender, and with very slender members) the issue of structural stability should also be addressed. Therefore, it is expected that the combination in the use of the information already available, with the procedure herein developed, together with the consideration of more advanced topics related to strength and structural stability evaluation, will bring a deep insight in the behavior of the towers.

ACKNOWLEDGMENTS

This study was funded in part by the Universities of Salamanca and Vigo. Authors would like to thank Remote Sensing and the anonymous reviewers which provided numerous comments and suggestions that contributed to improve significantly an earlier version of the manuscript.

AUTHOR CONTRIBUTIONS

All authors contributed extensively to the work presented in this paper.

Conflicts of Interest

The authors declare no conflict of interest.

REFERENCES

1. Ghilani, C.D.; Wolf, P.R. Adjustment Computations: Spatial Data Analysis, 4th ed.; John Wiley & Sons: Hoboken, NJ, USA, 2006.

2. Villarino, A.; Riveiro, B.; Gonzalez-Aguilera, D.; Sanchez-Aparicio, L. The integration of geotechnologies in the evaluation of a wine cellar structure through the Finite Element Method. Remote Sens. **2014**, 6, 11107–11126.

3. Costanzo, A.; Minasi, M.; Casula, G.; Musacchio, M.; Buongiorno, M.F. Combined use of terrestrial laser scanning and IR thermography applied to a historical building. Sensors **2014**, 15, 194–213.

4. Fregonese, L.; Barbieri, G.; Biolzi, L.; Bocciarelli, M.; Frigeri, A.; Taffurelli, L. Surveying and monitoring for vulnerability assessment of an ancient building. Sensors **2013**, 13, 9747–9773.

5. Choi, S.W.; Kim, B.R.; Lee, H.M.; Kim, Y.; Park, H.S. A deformed shape monitoring model for building structures based on a 2D laser scanner. Sensors **2013**, 13, 6746–6758.

6. Ioannidis, C.; Valani, A.; Georgopoulos, A.; Tsiligiris, E. 3D model generation for deformation analysis using laser scanning data of a Cooling Tower. In Proceedings of the 3rd IAG/12th FIG Symposium, Baden, Switzerland, 22–24 May 2006.

7. González-Aguilera, D.; del Pozo, S.; Lopez, G.; Rodriguez-Gonzálvez, P. From point cloud to CAD models: Laser and optics geotechnology for the design of electrical substations. Opt. Laser Technol. **2012**, 44, 1384–1392.

8. Shu, Q.; Yuan, G.; Zhang, Y.; Guo, G. Research on anti-foundation-displacement performance and reliability assessment of 500 KV transmission tower in mining subsidence area. Open Civ. Eng. J. **2011**, 5, 87–92.

9. Yang, F.; Yang, J.; Han, J.; Zhang, Z. Study on the limited values of foundation deformation for a typical UHV transmission tower. IEEE Trans. Power Deliv. **2010**, 2, 2752–2758.

10. Yuan, G.L.; Li, S.M.; Xu, G.A.; Si, W.; Zhang, Y.F.; Shu, Q.J. The anti-deformation performance of composite foundation of transmission tower in mining subsidence area. Proc. Earth Plan. Sci. **2009**, 1, 571–576.

11. Shu, Q.J.; Yuan, G.L.; Guo, G.L.; Zhang, Y.F. Limits to foundation displacement of an extra high voltage transmission tower in a mining subsidence area. Int. J. Min. Sci. Technol. **2012**, 22, 13–18.

12. Yang, F.; Li, Q.; Yang, J.; Zhu, B. Assessment on the stress state and the maintenance schemes of the transmission tower above goaf of coal mine. Eng. Fail. Anal. **2013**, 31, 236–247.

13. Prasad Rao, N.; Kalyanaraman, V. Non-linear behaviour of lattice panel of angle towers. J. Constr. Steel Res. **2001**, 57, 1337–1357.

14. Da Silva, J.G.S.; da S. Vellasco, P.C.G.; de Andrade, S.A.L.; de Oliveira, M.I.R. Structural assessment of current steel design models for transmission and telecommunication towers. J. Constr. Steel Res. **2005**, 61, 1108–1134.

15. Prasad Rao, N.; Samuel Knight, G.M.; Mohan, S.J.; Lakshmanan, N. Studies on failure of transmission line towers in testing. Eng. Struct. **2012**, 35, 55–70.

16. Klinger, C.; Mehdianpour, M.; Klingbeil, D.; Bettge, D.; Häcker, R.; Baer, W. Failure analysis on collapsed towers of overhead electrical lines in the region Münsterland (Germany) 2005. Eng. Fail. Anal. **2011**, 18, 1873–1883.

17. Zhuge, Y.; Mills, J.E.; Ma, X. Modelling of steel lattice tower angle legs reinforced for increased load capacity. Eng. Struct. **2012**, 43, 160–168.

18. Órgano MINISTERIO DE INDUSTRIA, TURISMO Y COMERCIO. Real Decreto 223/2008 Reglamento sobre Condiciones Técnicas y Garantías de Seguridad en Líneas Eléctricas de Alta Tensión y sus Instrucciones Técnicas Complementarias ITC-LAT 01 A 09, 1st ed.; Boletín Oficial del Estado: Madrid, Spain, 2008.

19. Wilson, E. Integrated Finite Element Analysis and Design of Structures, 7th ed.; Computers and Structures, Inc: Berkeley, CA, USA, 1998.

20. Eurocode 3: Design of steel structures—Part 1-1: General Rules and Rules for Buildings EN 1993-1-1: 2005. Available online: https://law.resource.org/pub/eur/ibr/en.1993.1.1.2005.pdf (accessed on 4 July 2015).

21. Beer, F.; Johnston, E.R., Jr.; Mazurek, D. Vector Mechanics for Engineers: Statics, 10th ed.; McGraw-Hill Science: London, UK, 2012.

22. Hibbeler, R.C. Mecánica Vectorial Para Ingenieros: Estática, 10th ed.; Pearson Educación: Mexico D.F., México, 2004.

23. Eurocode 1: Actions on structures Part 1-1: General Actions—Densities, Self-Weight, Imposed loads for buildings. UNE-EN 1991-1-1: 2003. Available online: http://infostore.saiglobal.com/emea/details. aspx? ProductID=400059 (accessed on 4 July 2015).

24. Argüelles, R.; Arriaga, F.; Atienza, J.R.; Martinez, J.J. Estructuras de Acero II: Uniones y Sistemas Estructurales, 1st ed.; Bellisco: Madrid, Spain, 2001.

25. Besl, P.; McKay, N. A method for registration of 3D shapes. IEEE Pattern Anal. Machine Intell. **1992**, 14, 239–256.

26. Eurocode 0—Basis of Structural Design UNE-EN 1990: 2003. Available online: http://infostore.saiglobal.com/store/details.aspx?ProductID=400051 (accessed on 4 July 2015).

27. Laefer, D.F.; Truong-Hong, L.; Fitztgerald, M. Processing of terrestrial laser scanning point cloud data for computational modelling of building facades. Recent Pat. Comput. Sci. **2011**, 44, 16–29. [

28. Liu, Y.C.; Liu, Z.D. Study on stabilization and rectification technology for inclined transmission tower. Rock Soil Mech. **2008**, 29, 173–176.

29. Villarino, A.; Riveiro, B.; Martínez-Sánchez, J.; Gonzalez-Aguilera, D. Successful applications of geotechnologies for the evaluation of road infrastructures. Remote Sens. **2014**, 6, 7800–7818.

CHAPTER 8

Automatic Extraction of Railroad Centerlines from Mobile Laser Scanning Data

Sander Oude Elberink [1,*] and Kourosh Khoshelham [1,2]

[1] Faculty of Geo-Information Science and Earth Observation, University of Twente, P.O. Box 217, 7514 AE Enschede, The Netherlands

[2] Department of Infrastructure Engineering, University of Melbourne, Melbourne 3010, Australia

ABSTRACT

In this paper, we describe the automatic extraction of centerlines of railroads. Mobile Laser Scanning systems are able to capture the 3D environment of the rail tracks with a high level of detail. Our approach first detects laser points that were reflected by the rail tracks, by making use of local properties such as parallelism and height in relation to neighboring objects. In the modeling stage, we present two approaches to determine the centerline location. The first approach generates center points in a data-driven manner by projecting rail track points to the parallel track, and taking the midpoint as initial center point. Next, a piecewise linear function is fitted through the center points to generate center points at a regular interval. The second approach models the rail track by fitting piecewise 3D track models to the rail track points. The model consists of a pair of two parallel rail tracks. The fitted pieces are smoothened by a Fourier series interpolation function. After that the centerline is implicitly determined by the geometric center of the pair of tracks. Reference data has been used to analyze the quality of our results, confirming that the position of the centerlines can be determined with an accuracy of 2–3 cm.

KEYWORDS

mobile laser scanning; point clouds; railway; detection; centerline; 3D modeling; stochastic model fitting; rail tracks

1. INTRODUCTION

The centerline of a rail track is an important element at various stages in the object life of the rail track. In railroad designs, the centerline is used to determine features such as maximum curvature and impact on the environment. Required clearances near the railroad are based on the location of the centerline.

Rail track center lines are also an important data input for track alignment analysis. Rail track operators seek to achieve the most optimal curves to provide a smooth and safe ride for the passengers. Over time, rail tracks tend to shift, thereby degrading the optimal rail path. If the current rail geometry is known, the difference between the optimal path and the actual situation can be computed. Huge machines, such as ballast tempers are used to shift and tilt the rail back into the optimal shape.

The input data for the alignment analysis requires very high accuracy. Traditionally, manual surveys using tachymeters were needed to obtain the data. Currently, mobile mapping systems (MMS) are capable of recording the railroad environment with similar accuracy as manual surveys. Most MMS are equipped with positioning sensors, such as GPS and INS sensors, and image and range sensors, such as cameras and mobile laser scanning (MLS) systems. A broad range of mobile mapping systems is on the market, as described by [1,2]. Although most of the systems are mounted on a car acquiring street level objects, MMS systems can also be mounted on other platforms such as trains and boats [3]. The main difference with manual surveys is that the recorded data is an uninterpreted image or point cloud, instead of a labelled set of interpreted data points.

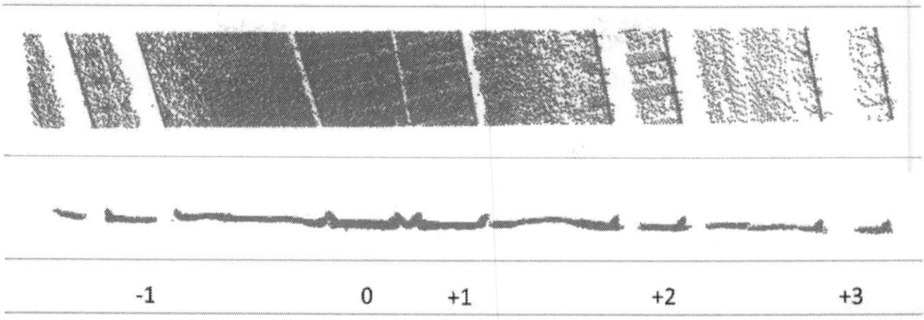

Figure 1. Varying point density across neighboring rail tracks. Top view (**top**), cross section (**middle**), and rail track number (**bottom**). The mobile mapping system was mounted on a platform driving at track “0”. It is our aim to find the centerlines of all tracks.

The point cloud data can be used for rail track and wires detection, asset management of poles, signs and towers. The promising results on 3D modeling of rail tracks [4] raised the question whether it would be possible to automatically extract the centerline of railroads, with an accuracy better than 1.5 cm standard deviation which is considered to be the accuracy of current semi-automatic or manual measurements. Additional question was if centerline generation is also possible on data where the MMS system is located at a track next to the track of interest. This is of practical importance as it is fairly impossible to drive at every single track of a whole country within a given timeframe without interfering with the regular train schedule. In addition, tracks may be inaccessible when other trains are blocking the entrance. As shown in Figure 1, the distribution of points over the different tracks is far from constant. Average point densities at the track bed quickly decrease from 5000 p/m^2 in track “0” to 500 p/m^2 in track -1, to less than 100 p/m^2 in track “+2” and about 40 p/m^2 in track “+3”. In addition, clearly visible is the existence of large data gaps due to occlusions by the rail track itself.

In this paper, we present two approaches to generate center lines of rail tracks. Both are based on a rail track extraction algorithm as a first step. The first approach is based on a direct calculation using detected rail track points. The second approach generates 3D models per pair of rails by fitting parametric models to the detected rail track points, where implicitly the center lines are determined by the origins of these fitted models. It is of interest to analyze which approach gives more accurate results, in order to propose an alternative to existing semi-automated approaches.

The contribution of this paper is two-fold:

- To compare a data-driven and a model-driven approach to centerline extraction;
- To evaluate the accuracy of centerlines obtained by the two methods.

After a literature review in Section 2, the methodology is described in Section 3. Results are shown in Section 4, followed by the conclusions in Section 5.

2. LITERATURE REVIEW

In the past, several researchers have focused on object extraction at or near the railway environment. This section on related literature first focusses on object detection near rail tracks, followed by a paragraph on object modeling and curve fitting approaches.

Kaleli and Akgul [5] efficiently made use of the constant forward viewing angle of cameras mounted on a locomotive to determine the left and right rail track. The assumption was that the acquisition platform was at the same track as the objects of interest. Zhu and Hyyppa [6] focused on the railroad environment using a combination of mobile and airborne laser data. Their objects of interest included towers, poles, terrain, trees and nearby building facades and roofs. Modeling of the rail track itself was not part of their focus. Beger et al. [7] has been one of the rare papers on calculation of the center lines using airborne laser scanner data in combination with high resolution imagery, based on earlier work of [8]. They have used imagery to mask the rail tracks in the laser data. Next, tracks are detected and center lines have been determined. Although geometric accuracy results have not been reported, the completeness results indicated that the vast majority of center points felt within 15 cm of reference centerlines. Oude Elberink et al. [4] has described a preliminary method to detect and model rail tracks in MLS data based on local and global properties of the rail environment. The detection rate for the classified points was about 95%, which seems to be more than enough for modeling the rail tracks. A limitation of that approach was the inability to detect track points near switches, and also no efforts were made to determine the centerline. Yang and Fang [9] have extracted rail tracks in MLS data by combining geometric and radiometric information. Using mainly a data-driven approach they have reached an overall detection rate of about 95%. The disadvantage of such data-driven approach is the direct relation between point density and the detection rate.

The authors also concluded that at switch locations the correctness of extraction is lower than for single road tracks.

Fitting parametric models to point clouds is a well-known processing step in 3D modeling, for which many techniques have been used [10]. Examples can be found for building models [11,12], industrial installations [13], and indoor scenes [14]. Next to fitting a parametric model to the detected rail points, also a curve fitting step is applied to the fitted models in order to ensure a smooth rail track model. In [15], a Bezier curve fitting approach has been applied, which was improved for more curvy situations in [4] by a Fourier series curve fitting. Curve fitting approaches were used in [16,17] to calculate accurate road center lines by smoothening detected segments. [18] has shown the use of static Terrestrial Laser Scanning (TLS) data for fitting predefined rail track models for monitoring purposes at train stations. The quality of their data was extremely high, making it possible to fit 3D models with an accuracy of better than 3 mm relative to the point cloud.

This contribution contains the description of the detailed determination of center lines of rail tracks from mobile laser scanning data, even if the data is acquired from a sensor system at a neighboring rail track. Compared to [7], our results should be more detailed because of the higher density data, inclusions of a model fitting step to bridge data gaps as well as to reduce the influence of data noise.

3. METHODOLOGY OF CENTER LINE EXTRACTION

3.1. Overview of the Methods

The centerline of a rail way track is an imaginary line in between two parallel rails, as shown in Figure 2. In the terminology of this paper, a track consists of a pair of rails.

MLS data delivers a dense description of the rail way environment, represented by hundreds or even thousands of 3D points per m^2. Our method is designed to automatically reconstruct the position of the centerline using two approaches, as shown in Figure 3. They both start by detecting the points that hit the rail way tracks. The first is a data driven approach that calculates imaginary points in between points on two parallel rail tracks, followed by a curve fitting algorithm to get a smooth centerline. The second approach generates a 3D model of the rail tracks, taking into account parallelism and constraints to the smoothness and slope of rail way design.

Based on the location of the 3D models, the centerline is implicitly defined as the line in the middle of the pair of rails.

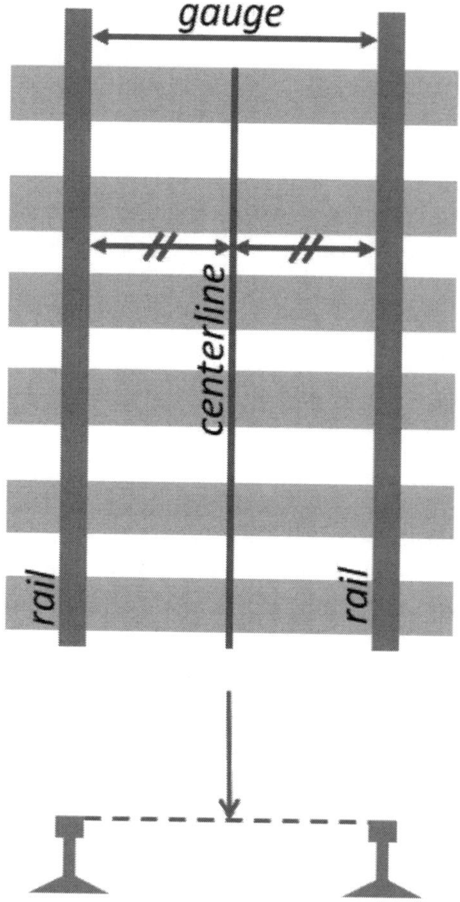

Figure 2. Schematic presentation of centerline, gauge and rails.

3.2. Initial Rail Track Detection by Grid Wise Height Histogram Analyses

Generally, the rail track environment is an extremely standardized area, in terms of types and shapes of objects, open spaces, and distances between objects. This knowledge is integrated in the design of our method to detect and model rail tracks. For finding the majority of rail tracks, it can be assumed that the track itself is at or near the ground, that the space just above the tracks is empty, and that there may be contact wires and other wires at a certain height above the track. For detecting rail tracks in a mobile laser scanning point cloud, this can be implemented by checking whether these elements are presents in a certain area. We perform a coarse track detection by analyzing the height histogram of laser points in a 1 by 1 m neighborhood. In the lowest 0.5 m a search for points on a rail track is performed if the space above is rather empty. The height histogram analysis starts by taking the 10-th percentile point within each 1 × 1 m grid as DTM height. Next, if there are rails in this 1 square meter, there should be a rather empty space in between 0.5 and 4.5 m above this DTM height. If that is the case, then the terrain points are further analyzed by checking whether there are a few points that are slightly higher than the other terrain points. These points are initially classified as rail track points.

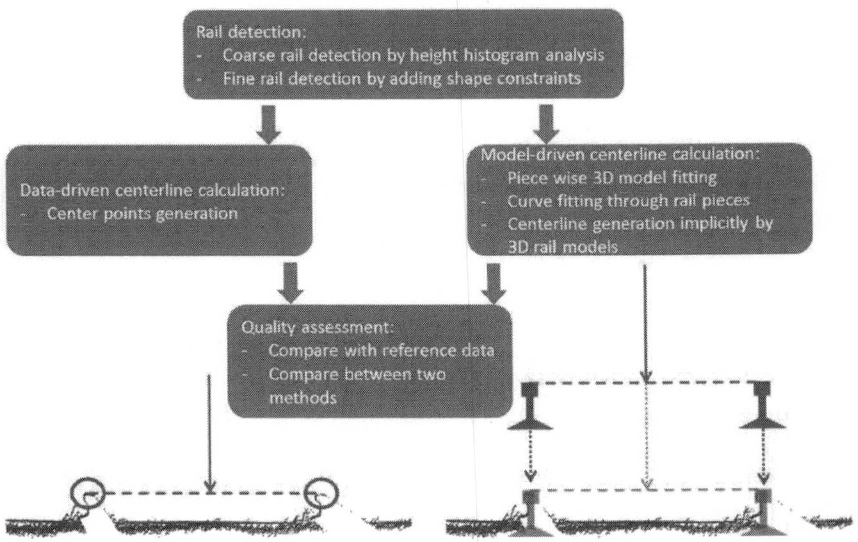

Figure 3. Schematic workflow from rail detection to data-driven and model-driven centerline calculation.

The result of this step is a coarse indication which points are at the rail track, based on the relative height to the ground and potential wires. There are situations where points may be either incorrectly detected as rail track (false positives) or not detected (false negatives). In order to remove false positives and to reduce the number of false negatives, we incorporate more shape knowledge of the rail track in the fine detection stage. A preliminary approach was described in [4]. One of the limitations of that approach was the implication of the RANdom Sample Consensus (RANSAC) line fitting approach which selects the best fitted line within a certain local neighborhood. As the neighborhood was defined by a regular grid of 1 × 1 m, only one rail track could be found within every m^2: this meant that at switches part of the points on tracks were not selected. In this paper, we directly calculate point based features without the need to keep only the RANSAC inliers of a local fitted line. As a direct consequence, points on tracks near switches can still be detected automatically. As the point-based feature calculation is relatively expensive, the coarse track detection applied in this step helps reducing the computation time by eliminating the places where definitively no rail tracks are present.

3.3. Fine Rail Track Detection by Point Based Attribute Calculation

In this refinement stage, shape knowledge is used that a rail track locally is linear and two rails of a track are parallel. Therefore, we assume that points on rail tracks also show this pattern of linearity and parallelism. Per point the direction of nearby initially detected rail points is calculated, followed by an attribute that counts how many points are parallel at a given gauge distance.

For calculating the direction attribute for every coarsely detected rail point, nearby detected rail track points have been selected, e.g., within 1 m radius. Per point, the line is found iteratively such that most of the neighboring points is passing through that line within 3 cm distance. The line direction is taken as the direction attribute. Next, for every rail point we construct a parallel buffer at the gauge distance, and count the number of rail points that fall inside that buffer. Additional condition is that those points should have a similar direction value as well. If the number of points in the parallel buffer exceeds a threshold value, the point is classified as refined rail track point.

Due to the scanning configuration, it cannot be assumed that the points are symmetrically distributed over the tracks and over the rails per track. One side of a rail may be occluded for the scanner. This may have an influence on the determination of the centerline: the distribution of points may lean slightly towards the position of the scanner, causing that the centerline is not exactly in the middle of two point clouds of parallel tracks. Two different approaches are presented in 3.4 and 3.5 and aim to minimize the influence of this possibly asymmetric distribution of points.

3.4. Centerline Determination Directly from Detected Points

The data-driven approach selects only the top points of the detected tracks, assuming that the asymmetry in distribution is less systematic, even almost absent, at the top of the rail. For every point, the local direction of the track is known from the "direction" attribute, and it is known which points are at the parallel track within a certain radius. A line is fitted to points at the parallel track which contain the same direction value. Next, the perpendicular direction is determined and used to construct an across track line projected through the processed point. The midpoint of that projection is stored in a vector of center points, see Figure 4. The consequence is that for every detected rail track point one center point is generated, based on fitting a line through points at the parallel track. Additional property is that the center points are determined by parallel points with a distance of the gauge ±10 cm in-between the rail points. This means that even at switch locations center points have been determined by a pair of corresponding rail track pieces. The switch itself is recognized automatically as centerlines of one track will approach and even cross the rail track points of the track that joins or split from the first track. In our implementation, a minimum threshold of 0.6 m between center points and rail track points is used for single tracks, as by definition the distance should be around half the gauge distance, i.e., 0.70–0.75 m. If center points and rail tracks points are within 0.6 m, this can only mean a situation where the centerline is from another track than the track points, as shown in Figure 5.

Next, points are grouped into segments. This holds for both the center points and the rail track points. The working of the segmentation is based on component growing algorithm using the maximum distances between points, constraining the point attributes of the local direction and their label, e.g., switch, left/right track and center points. Data gaps in along-track direction can be covered now that we can locally calculate the direction of the rail way. This enables us to group collinear segments. We limit the maximum data gap to cover to 8 m in track direction, in order to avoid possible incorrect grouping of different rail ways. At switch locations, our algorithm will stop the component growing as during the modeling stage we restrict to model parts in between two switches to reduce the degree of freedom. Rail track points and their corresponding center points at switch locations are grouped in a separate connected component. The result after this step is a labelled and segmented point cloud, where the segments correspond to groups of points on normal rail tracks, center points or switch points.

The generated center points have been used as input for an interpolation algorithm for our first approach to reconstruct the centerlines. Per rail track, all center points are collected and split into regular intervals, e.g., 2.0 m, along the rail track. For each interval one position is stored at the middle location of a straight RANSAC line fitted through the majority of points in that interval. If an interval appears to have no center points due to a local data gap, center points are interpolated at a straight line from the previous known center location until the next interval after the gap. The second approach for center line reconstruction, discussed in the next section, is based on 3D rail track modeling, making use of the detected rail track points and a parametric track model.

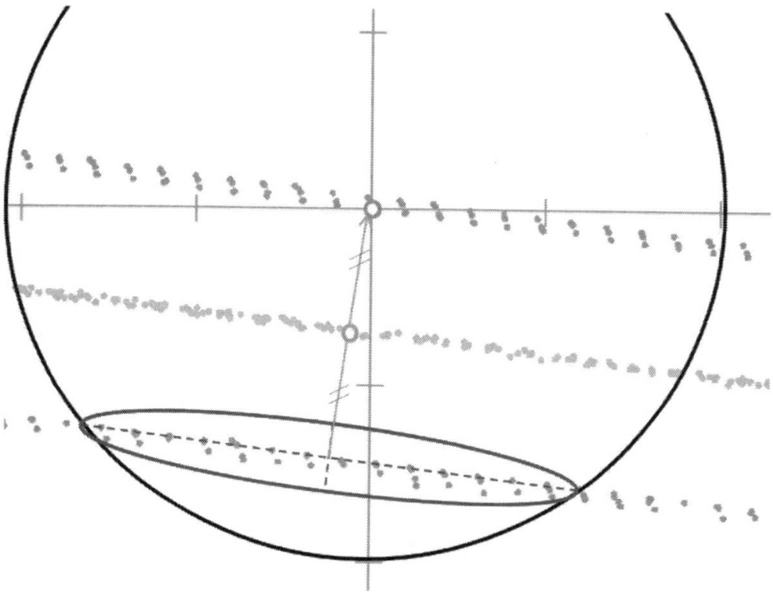

Figure 4. Within a radius of 2 m (black circle), points are selected at the parallel track (red ellipse). The average direction of those points is used to construct a perpendicular line through the current rail track point (green dot). The center point is generated on that line (blue dot), exactly in between the rail track point and line through the selected parallel track points.

3.5. Centerline Determination by 3D Modeling of the Rail Tracks

The model-driven approach calculates 3D models by fitting parametric track models to the detected rail track points. For this approach not only the top of the track is considered but also the foot of the track. For this, all points directly below the detected rail top are included in the detected rail track data as well. The 3D rail model is generated by fitting parametric pieces to point segments along the rail, followed by interpolating a smooth and continuous rail model between the pieces. The following sections describe the steps in this process.

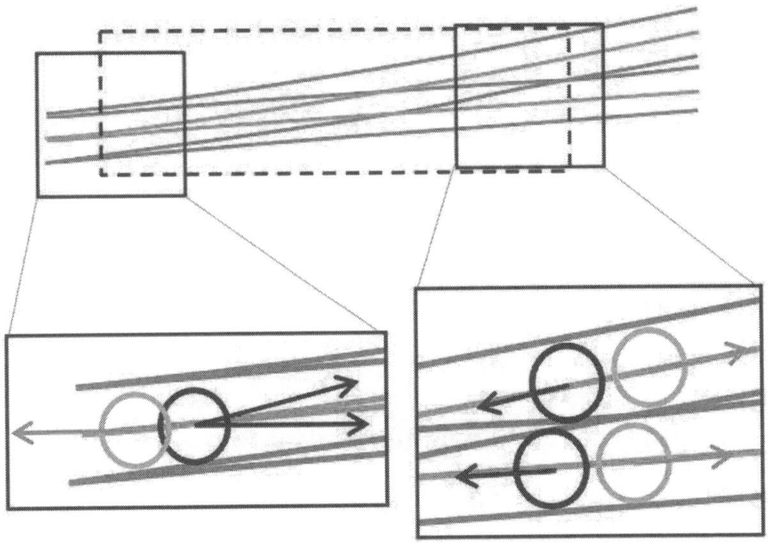

Figure 5. Switch detection (dashed red box) by checking distances between rail track points and center points. Red circles indicate locations where rail track points are closer than e.g., 60 cm to the center points. Grey circles indicate normal situations where center points and track points are more than 60 cm apart.

3.5.1. Defining a Parametric Model of a Rail Piece

The parametric model consists of a pair of two simplified rail pieces, as shown in Figure 6. It is defined by eight shape parameters and six orientation parameters. The shape parameters are considered fixed throughout the track, and are obtained from standard railway specifications. The orientation parameters define the position and rotations of the local coordinate system of the model with respect to the global coordinate system of the point cloud, and are estimated for each piece. Compared to our previous model [4], where the left and the right rail were modelled independently, the current model allows modeling both rails simultaneously. This guarantees that the left and right pieces always have the same orientation, and that the final model has a constant gauge distance throughout the rail track.

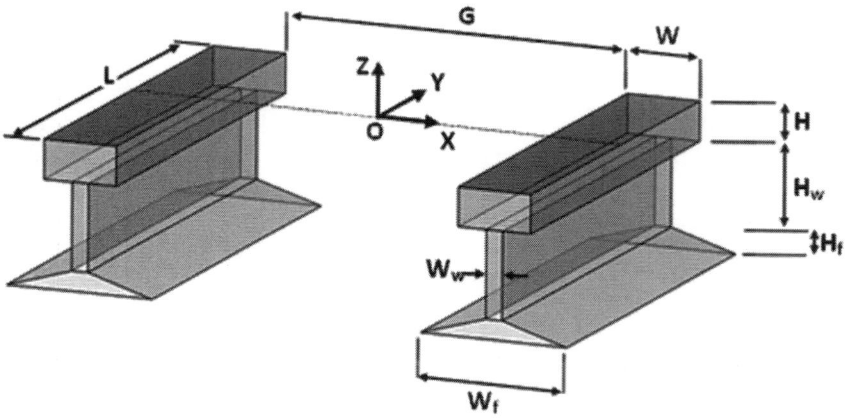

Figure 6. Parametric model of a pair of two rail pieces. Position and the orientation parameters of the local coordinate system need to be determined.

3.5.2. Estimating the Parameters of a Piece Model

To fit the parametric pieces to the points, first the points detected on both rails are partitioned into small segments of size L (length of the model) along the main axis of the track. For this purpose, the center points generated in the previous step are used as the approximate axis of the track.

The common approach to parametric model fitting is by least-squares estimation of the model parameters, minimizing a measure of distance between the model and the data [19,20]. However, the least-squares estimation method is largely influenced by the missing points and outliers, which are difficult to avoid in the detection step. For this reason, we adopt a Markov Chain Monte Carlo (MCMC) algorithm to estimate the parameters of each piece by sampling the joint probability distribution of the orientation parameters. Formally, for a given set of points X_j in segment j, the aim is to find a model Θ_j that maximizes the probability $p(\Theta_j|X_j)$, where $\Theta_j = [x_{oj}, y_{oj}, z_{oj}, \omega_j, \phi_j, \kappa_j]^T$ is the vector of six orientation parameters of the model. Using Bayes' rule:

$$p(\Theta_j|X_j) = \eta p(X_j|\Theta_j)p(\Theta_j) \tag{1}$$

where $p(X_j|\Theta_j)$ is the evidence for the model provided by the points, $p(\Theta_j)$ is the model prior, and $\eta=1/\sum\Theta i p(Xj|| \Theta i)p(\Theta i)$ is a normalization factor that does not affect the choice of Θ_j. We define the evidence as:

$$p(X_j|\theta_i) = \exp(-\frac{\bar{d}_j^{\,2}}{c^2}) \qquad (2)$$

where $\bar{d}j$ is the mean distance between the points in segment j and the model Θ_j , and c is a constant. The point-model distance is defined as the perpendicular distance from a point to its corresponding planar patch in the model. To find the corresponding patch, the point is orthogonally projected to each of the patches of the model, and the projected point is tested whether it is contained within the boundary of the patch. The patch that contains the orthogonal projection of the point is selected as the corresponding patch, and its perpendicular distance from the point is calculated. Points that do not project to any of the patches (these are usually points that are far away from the model) are excluded from the calculation of $\bar{d}j$ in Equation (2).

The prior $p(\Theta_j)$ is used to incorporate our prior knowledge of the model parameters. We expect that the rotation parameters of each rail piece are only slightly different from the previous piece, and that the position of a piece is close to the center of its corresponding point segment. To include this knowledge, we define the prior as a normal distribution centered around the expected rotation and position parameters of the piece model:

$$p(\Theta_j) \sim N(\mu_j, \Sigma_j) \qquad (3)$$

where μ_j is the mean vector and $\sum_j$ is the covariance of the parameters in Θ_j. The mean vector contains the expected rotation (i.e., the rotation of the previous piece) and position (i.e., the center of the point segment) of the piece. For the first piece in each track we set $\omega_1 = 0$ and $\phi_1 = 0$, while κ is obtained as the direction of the approximate centerline. The covariance of the prior distribution $\sum_j$ is defined by assuming large variances for the expected orientation parameters and no correlation between them. The large variances ensure that the Markov chain does not get stuck at the mean of the prior distribution.

For the MCMC sampling the Metropolis-Hastings algorithm [21] is used. The algorithm starts at a random initial point and recursively provides samples Θ_t, t = 1,…,n from the target distribution $p(\Theta_j|X_j)$. The samples are drawn from a proposal distribution, which is chosen as a Gaussian with variances larger than the prior distribution. An estimate of the expectation of model parameters is then obtained by the ergodic mean of the samples: *θ^j=1n−m∑nt=mθt*, where m specifies the number of so-called burn-in samples. Further details about the Metropolis-Hastings algorithm can be found in [22].

3.5.3. Interpolating a Smooth and Continuous Rail Model

The rail pieces obtained by the MCMC sampling do not necessarily form a continuous and smooth rail track. The estimated orientation parameters are defined at the local origins (of the coordinate systems of the pieces) at intervals of approximately L (length of a piece). To obtain a continuous and smooth model of the rail tracks, the orientation parameters should be interpolated in between these local origins. As rail tracks typically consist of straight segments and circular curves, we choose a Fourier series as the interpolation function:

$$\theta^i(t) = \sum_{k=0}^{n} \left[a_k \cos(kwt) + b_k \sin(kwt)\right] \tag{4}$$

where $\Theta^i(t)$, i = 1,..,6 is i-th orientation parameter at an intermediate location t in the interval [1, m], m being the number of rail pieces, a_k, b_k, w are the unknown coefficients of the interpolation function and n is its order. The interpolation coefficients are estimated for each parameter separately. By evaluating Equation (4) with the known Θ^i of all pieces, a system of equations is obtained, which is then solved for the unknown coefficients. The estimated coefficients minimize the sum of squared differences between the piece parameter and the interpolated value. The order of the Fourier series defines the flexibility of the fitted curve. However, higher order Fourier series may result in oscillation and waviness in the final model. By experiment we found that the third order Fourier series is sufficient to fit a smooth curve to the orientation parameters accurately.

Once the interpolation coefficients are estimated for each orientation parameter, we can reconstruct a continuous and smooth model of the rail track by connecting very small rail pieces using the interpolated orientation parameters. The centerline of the reconstructed rail track is simply obtained as the set of interpolated position parameters along the rail track.

4. RESULTS

4.1. Dataset Description

To evaluate the methods, MLS data were acquired in two study areas. The first dataset was acquired at the Blerick train station near Venlo, and the second dataset was acquired at the train station of Sittard, both cities located in the south of the Netherlands. In Figure 7, laser data of two areas are shown. The data has been captured by Fugro using the RAIL-MAP system. This system contains a GNSS/INS system, a panoramic camera and two laser scanners. The system is dedicated and certified for use on railways. It can be mounted on a train and operate at regular track speeds, such that the train schedule will not be disrupted. The scanners are based on the measurement of phase shift. Each scanner is capable of registering up to 1 million points per second. Both scanners send the laser wave in a slightly inclined plane to reduce shadowing effects. The range accuracy is in the order of 1–3 mm, whereas the positional accuracy of the system as a whole is in the order of 1 cm. All areas contain several parallel rail tracks. Green arrows indicate the tracks at which the system has acquired the data. In Figure 7 area "Blerick" is shown, containing at least 6 parallel rail tracks where the system was mounted on a vehicle on the rightmost rail track. The rightmost track is at an across distance of between 25 and 30 m from the left track. One switch is located in the right part of this dataset. This size of the area is 350 × 40 m, containing 5.8 Million points. In Figure 7 the dataset of "Sittard", 600 × 80 m containing about 25 Million points, is shown with 9 parallel rail tracks on which on two of the tracks the MMS-vehicle has driven to acquire data.

Reference data has been used for testing the positional accuracy of the extracted centerlines. This reference data has been captured by Fugro with the RiLa system. This is a system carrier two laser scanners recording only the tracks directly below the sensor (RiLa, 2015), determining track geometry with an absolute accuracy of 1.5 cm. For the reference data of Sittard it was known at forehand that some of the reference centerlines would have an unknown systematic offset of a few centimeters. For Sittard we therefore calculate and eliminate the systematic error per rail track and focus on the standard deviation to the reference data.

Figure 7. Point cloud from area near station Blerick (**top**) and Sittard (**bottom**). Data has been acquired on tracks indicated by the green arrows.

4.2. Rail Track Detection

In Figure 8, part of the Blerick area is shown where points in green are roughly classified as rail track points. The points colored by segment in the lower figure are the refined rail track classification results. It can be seen that all of the false positives (indicated by red arrows) have been removed in the refinement step. The downside of this strict removal is that also some of the true positives are removed when the results become uncertain. This can be seen on the fourth parallel track, highlighted in the red ellipse. Three of the parallel tracks can be identified, plus the track where the system was driving. On the fourth parallel track, classification results become noisier and the refined classification step is not able to reliably detect points anymore. The reason is that we assume the rails are slightly elevated above the ground whereas on those rails there are no points on the ground; therefore rail points have not been detected. The parallel rail track can never be found in this case, when points from one of the two parallel rails have not been extracted in the rough classification. Although this may be a limitation of our algorithm, it is not considered to be a practical problem. In practice there will be enough possibilities to capture data on at least one of the other rail tracks instead of only the outer left one. An alternative solution is to integrate Airborne Laser Scanning (ALS) data to the MLS data to get the position of the terrain at the occluded areas in the MLS data. Once the terrain height is known from the ALS data, the detailed position of the rail tracks will come from the MLS points on the rail track.

Figure 8. Coarse (**top**) and fine (**bottom**) rail track detection in Blerick. MLS data has been captured at the most right rail track.

In the following figures, results of track detection and center line generation are shown. The upper row views of Figure 9 are constructed by an orthogonal projection of the points to a tilted plane, the bottom row are perspective views on a part of the data. It can be seen in the bottom row that center points can be generated even at locations where no rail points were detected on a path crossing the rail tracks.

Our second approach models 3D rail tracks by piece-wise fitting of parametric models followed by interpolation. The origins of the fitted pieces are the locations of the center lines. As a side-result, the 3D models are also generated. Results of the models are shown in Figure 10, together with the original point cloud and the locations of the generated center points. Rail models are represented by a wireframe model. Positions and orientations of the rail models are determined every 20 cm along the curve. Therefore, also the center points have been generated at intervals of 20 cm.

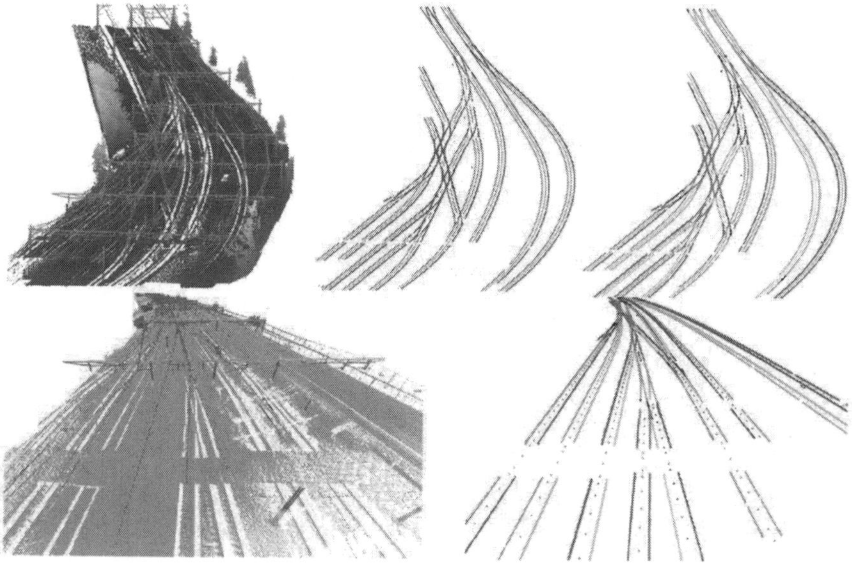

Figure 9. Rail track detection and centerline extraction from Sittard area. Top and bottom left: original point cloud; top center: points labelled by left/right rail piece, center points and switches; top and bottom right: points colored by connected component, center points derived at intervals of 2 m.

4.3. Accuracy Analysis of Center Line Extraction

Each generated center point is perpendicularly projected onto the corresponding reference centerline. In order to analyze whether potential discrepancies have a stochastic or systematic appearance, the signed distance has been calculated. Per rail track we analyze the behavior of the residuals between generated and reference center lines, by calculating the root means square error (RMSE), the bias and standard deviation per track. This is done for both the data-driven approach and the model-driven approach, i.e., after rail track fitting. The reference data itself has an expected standard deviation of 1.5 cm per center point. For both of our approaches the results are visualized in Figure 11.

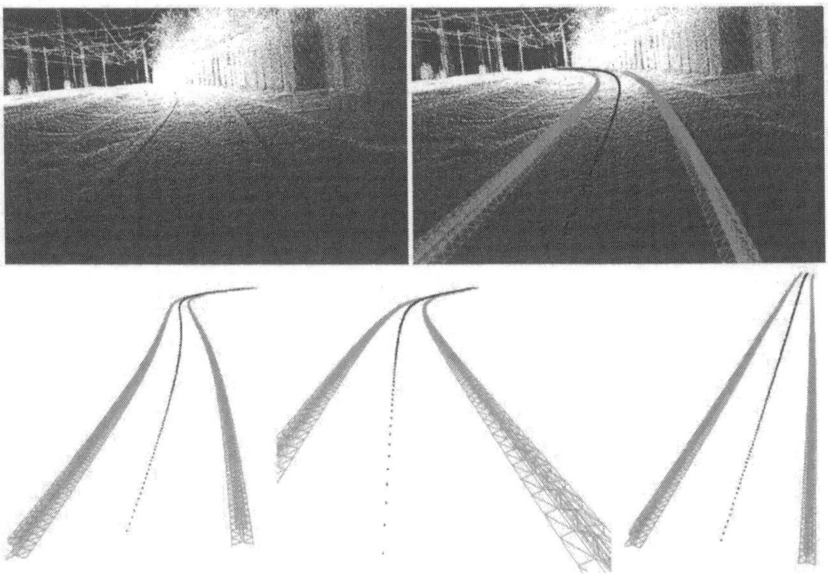

Figure 10. 3D Models and centerlines overlaid on raw point cloud data.

Point colors indicate the magnitude of the residuals at the generated center points. Grey lines represent the direction and magnitude of the residuals between generated center points and the corresponding reference centerline. The length of the lines is magnified by a factor 20 to enhance the interpretation of the figures.

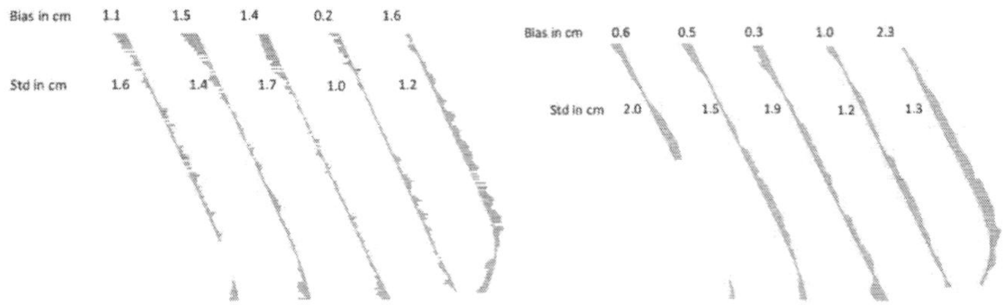

Figure 11. Residuals between generated centerlines and reference data for our data-driven approach (**left**) and the model-driven approach (**right**) in Blerick. Grey lines represent the direction and magnitude of the residuals (magnified by a factor 20).

Analysis of the behavior of the residuals in done in two directions:

- Along track direction; to analyze systematic patterns caused by curve fitting in our second approach and/or curve fitting in the reference data.
- Across track direction; to analyze the capability to generate center lines at parallel rail tracks next to the current rail track. Both systematic and stochastic errors are analyzed. The scanning angle may cause systematic effects for parallel tracks far away, and the decreasing point density may result in higher standard deviation values.

In along track direction one may see a small systematic pattern at the most right line, just after the curve for both methods in Figure 11. This is the lane where the system was driving. Residuals are between 2 and 3 cm at those locations, at other check points residuals are below 2 cm. The standard deviation of all residuals per track is listed in the figure varying from 1.0 to 2 cm. Note that in these values the effects of noise in the reference data is also included, meaning that both approaches have a standard deviation of better than 2 cm. In other words, the geometric quality of our results is comparable to the quality of the reference data.

In across track direction it can be seen that the standard deviation is increasing from 1.2 to 1.6 cm for approach 1 and 1.3–2 cm for approach 2 when moving further away from the sensor position. Although the pattern is not that strong, it is still arguable that the standard deviation increases from right to left. The bias does not show a clear pattern in across track direction. This means that our method is capable of generating centerlines at parallel tracks, with a slight decrease of precision for decreasing point densities.

For Sittard, the standard deviation of the residuals between reference data and generated center lines ranges from 1.3 to 2.7 cm, see Figure 12. The systematic offset (italic values in the figure) varies up to 5 cm per track. This can be caused by offsets in the reference data which have been produced on a track-wise base, or misalignments in the MLS system between position sensors and ranging sensors or a combination of these two causes. For Fugro this has been a reason to further analyze the chain of measurements in both the MLS system and the reference measurements to reduce the influence of potential systematic errors.

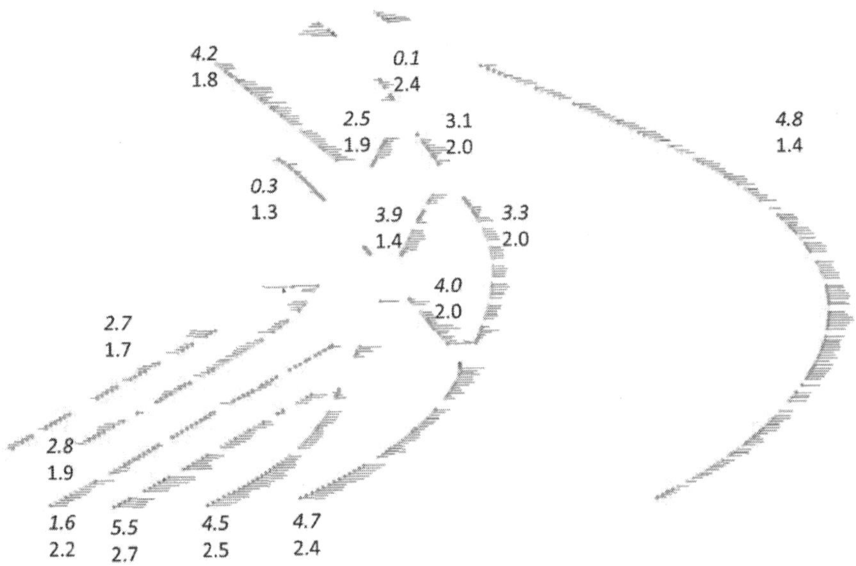

Figure 12. Residuals between centerlines from data-driven approach and reference data in Sittard: top value is the systematic offset, bottom value represent the standard deviation of the residuals per track. Grey lines represent the direction and magnitude of the residuals (magnified by a factor 20).

The influence of the modeling and Fourier series interpolation function is checked by comparing the center lines from the data-driven approach with the ones after 3D modeling and curve fitting. As the detected rail points are the same for both approaches, differences can only be explained by differences in the further processing steps. Differences are shown in Figure 13, where the maximum difference occurs at the most curved area in the datasets. The geometric differences of the center line location between the two approaches is mostly below 1 cm but can reach up to 2.7 cm in curved situations. This is a direct consequence of the use of the curve fitting algorithm, and how well the fitted curve fits to the rail track.

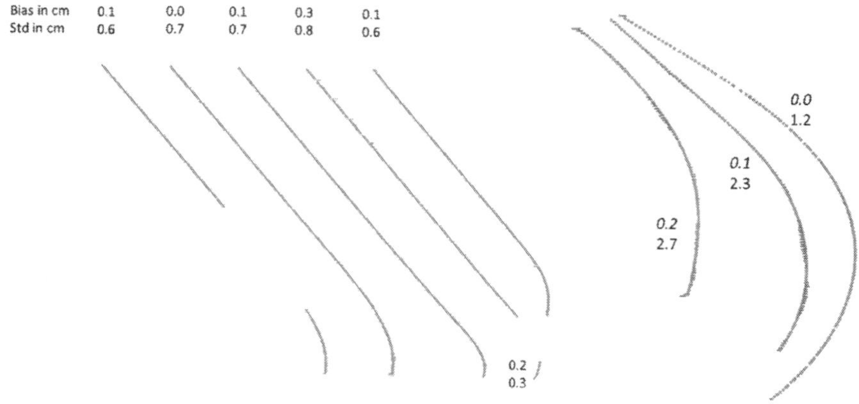

Figure 13. Differences between data-driven and model-driven approach in Blerick (**left**) and Sittard (**right**).

5. CONCLUSIONS

The fact that a pair of rail tracks is parallel and locally linear are the most essential properties that have been used in the rail track and center points' determination. These properties reduce the number of initially false detected rail points from the rough detection stage. Our approach can deal with local data gaps, as the scene constraints of rail way tracks are strong enough to bridge several meters without data. In cases of data gaps longer than 8 m at the rail track or its supporting ground surface, our algorithm cannot recognize connections between detected rail segments. In practice, it resulted in the correct reconstruction of four neighboring rail tracks, where the fifth contains too many data gaps due to self-occlusion by the rail track.

The aim of this research was to analyze whether we can accurately and automatically determine centerlines of rail tracks, even if the platform is at another rail track. We have shown that it is indeed possible to generate centerlines as long as rail tracks can be identified. Two approaches have been presented in this paper: the first determines the center position directly between points of a pair of rails. The second approach models the rail track by fitting a parametric model to the extracted rail points.

Both of our approaches prove that the geometric accuracy of the extracted centerlines is between 1 and 3 cm. The geometric differences of the center line location between the two approaches is mostly below 1 cm but can reach up to 2.7 cm in curved situations.

The geometric quality of the reference data is comparable with the quality of our results. Although this makes it challengeable to draw hard conclusions on the exact accuracy of our method, it proves that our approach is a very attractive alternative knowing that it is based on a fully automated procedure.

Future work includes automated extraction and modeling of other objects in a railway environment such as the platforms, catenary wires and towers. Next, existing 2D geo-information for the rail tracks will be updated and converted into 3D.

ACKNOWLEDGMENTS

The authors would like to deeply thank Martin Kodde and Rikkert Wienia from Fugro GeoServices not only for providing RAIL-MAP data and reference data, but also for their valuable input on the working of the systems and the practical importance of this method.

AUTHOR CONTRIBUTIONS

Next to the overall paper writing and organization, Sander Oude Elberink designed the method for rail point classification and data-driven center points generation. Kourosh Khoshelham performed the parametric modeling and curve fitting for the model-driven centerline extraction.

CONFLICTS OF INTEREST

The authors declare no conflict of interest.

REFERENCES

1. Petrie, G. Mobile mapping systems: An introduction to the technology. GeoInformatics **2010**, 13, 32–43.

2. Puente, I.; González-Jorge, H.; Martínez-Sánchez, J.; Arias, P. Review of mobile mapping and surveying technologies. Measurement **2013**, 46, 2127–2145.

3. Vaaja, M.; Kukko, A.; Kaartinen, H.; Kurkela, M.; Kasvi, E.; Flener, C.; Hyyppä, H.; Hyyppä, J.; Järvelä, J.; Alho, P.; et al. Data processing and quality evaluation of a boat-based mobile laser scanning system. Sensors **2013**, 13, 12497–12515.

4. Oude Elberink, S.; Khoshelham, K.; Arastounia, M.; Díaz Benito, D. Rail track detection and modelling in mobile laser scanner data. ISPRS Ann. Photogramm. Remote Sens. Spat. Inf. Sci. **2013**, 1, 223–228.

5. Kaleli, F.; Akgul, Y.S. Vision-based railroad track extraction using dynamic programming. In Proceedings of the 2009 IEEE 12th International Conference on Intelligent Transportation Systems, St. Louis, MO, USA, 4–7 October 2009; pp. 1–6.

6. Zhu, L.; Hyyppa, J. The use of airborne and mobile laser scanning for modeling railway environments in 3D. Remote Sens. **2014**, 6, 3075–3100.

7. Beger, R.; Gedrange, C.; Hecht, R.; Neubert, M. Data fusion of extremely high resolution aerial imagery and lidar data for automated railroad centre line reconstruction. ISPRS J. Photogramm. Remote Sens. **2011**, 66, S40–S51.

8. Neubert, M.; Hecht, R.; Gedrange, C.; Trommler, M.; Herold, H.; Krüger, T. Extraction of railroad objects from very high resolution helicopterborne lidar and ortho-image data. Int. Arch. Photogramm. Remote Sens. Spat. Inf. Sci. **2008**, 38, 25–30.

9. Yang, B.; Fang, L. Automated extraction of 3-d railway tracks from mobile laser scanning point clouds. IEEE J. Sel. Top. Appl. Earth Obs. Remote Sens. **2014**, 7, 4750–4761.

10. Tang, P.; Huber, D.; Akinci, B.; Lipman, R.; Lytle, A. Automatic reconstruction of as-built building information models from laser-scanned point clouds: A review of related techniques. Autom. Constr. **2010**, 19, 829–843.

11. Maas, H.-G.; Vosselman, G. Two algorithms for extracting building models from raw laser altimetry data. ISPRS J. Photogramm. Remote Sens. **1999**, 54, 153–163.

12. Verma, V.; Kumar, R.; Hsu, S. 3D building detection and modeling from aerial lidar data. In Proceedings of the 2006 IEEE Computer Society Conference on Computer Vision and Pattern Recognition, New York, NY, USA, 17–22 June 2006; pp. 2213–2220.

13. Rabbani, T.; Heuvel, F.V.D. Methods for fitting CSG models to point clouds and their comparison. In Proceedings of the 7th IASTED International Conference on Computer Graphics and Imaging, Kauai, HI, USA, 17–19 August 2004; pp. 279–284.

14. Lai, K.; Fox, D. Object recognition in 3d point clouds using web data and domain adaptation. Int. J. Robot. Res. **2010**, 29, 1019–1037.

15. Diaz Benito, D. Automatic 3d Modelling of Train Rails in a Lidar Point Cloud. M.Sc. Thesis, University of Twente, Enschede, The Netherlands, 2012.

16. Miao, Z.; Wang, B.; Shi, W.; Wu, H. A method for accurate road centerline extraction from a classified image. IEEE J. Sel. Top. Appl. Earth Obs. Remote Sens. **2014**, 7, 4762–4771.

17. Miao, Z.; Shi, W.; Zhang, H.; Wang, X. Road centerline extraction from high-resolution imagery based on shape features and multivariate adaptive regression splines. IEEE Geosci. Remote Sens. Lett. **2013**, 10, 583–587.

18. Soni, A.; Robson, S.; Gleeson, B. Extracting rail track geometry from static terrestrail laser scans for monitoring purposes. Int. Arch. Photogramm. Remote Sens. Spat. Inf. Sci. **2014**, XL-5, 553–557.

19. Lowe, D.G. Three-dimensional object recognition from single two-dimensional images. AI **1987**, 31, 355–395.

20. Khoshelham, K.; Li, Z.L. A model-based approach to semi-automated reconstruction of buildings from aerial images. Photogramm. Rec. **2004**, 19, 342–359.

21. Hastings, W.K. Monte carlo sampling methods using markov chains and their applications. Biometrika **1970**, 57, 97–109.

22. Gilks, W.R.; Richardson, S.; Spiegelhalter, D. Markov Chain Monte Carlo in Practice; Chapman and Hall/CRC: Boca Raton, FL, USA, 1995.

CHAPTER 9

Automatic Detection and Classification of Pole-Like Objects in Urban Point Cloud Data Using an Anomaly Detection Algorithm

Borja Rodríguez-Cuenca [1,*], Silverio García-Cortés [2], Celestino Ordóñez [2] and Maria C. Alonso [1]

[1] Department of Physics and Mathematics, University of Alcalá, Campus Universitario Ctra., Alcalá de Henares, 28871 Madrid, Spain

[2] Department of Mining Exploitation, University of Oviedo, Escuela Politécnica de Mieres, Gonzalo Gutiérrez Quirós, 33600 Mieres, Spain

ABSTRACT

Detecting and modeling urban furniture are of particular interest for urban management and the development of autonomous driving systems. This paper presents a novel method for detecting and classifying vertical urban objects and trees from unstructured three-dimensional mobile laser scanner (MLS) or terrestrial laser scanner (TLS) point cloud data. The method includes an automatic initial segmentation to remove the parts of the original cloud that are not of interest for detecting vertical objects, by means of a geometric index based on features of the point cloud. Vertical object detection is carried out through the Reed and Xiaoli (RX) anomaly detection algorithm applied to a pillar structure in which the point cloud was previously organized. A clustering algorithm is then used to classify the detected vertical elements as man-made poles or trees. The effectiveness of the proposed method was tested in two point clouds from heterogeneous street scenarios and measured by two different sensors. The results for the two test sites achieved detection rates higher than 96%; the classification accuracy was around 95%, and the completion quality of both procedures was 90%. Non-detected poles come from occlusions in the point cloud and low-height traffic signs; most misclassifications occurred in man-made poles adjacent to trees.

KEYWORDS

pole-like objects; feature extraction; pattern recognition; clustering; 3D point cloud; MLS; anomaly detection

1. INTRODUCTION

Creating and updating accurate maps and spatial databases has been demanded by various applications such as city management, urban planning, and intelligent transportation systems. For city management and urban planning, accurate land cover information is needed to document cities growth, make policy decisions, and improve land use planning [1]. For intelligent transportation systems, updated geodatabases that include the location of urban objects and traffic signs are required for terrestrial navigation and, of course, to decrease traffic congestion, lessen the risk of accidents [2], and develop autonomous driving systems [3]. Geospatial information has been widely used to meet these requirements for accurate and up-to-date remote sensed data. Light detection and ranging (LIDAR) technology has been used extensively in surveying and mapping. This technology provides three-dimensional data that complements the spectral information contained in two-dimensional images. Laser scanner sensors can be placed on aerial (airborne laser scanner, ALS) and terrestrial platforms (terrestrial LIDAR). Terrestrial LIDAR can be subdivided into two types: static and dynamic. Static terrestrial LIDAR technology (terrestrial laser scanner, TLS) data is collected from a sensor fixed in a base station. Thus, a small area can be mapped with high accuracy, but several scans are needed to cover large areas. Dynamic terrestrial LIDAR sensors (mobile laser scanner, MLS) are installed in vehicles provided with, as for ALS platforms, a navigation system based on global navigation satellite systems (GNSS) and inertial measurement units (IMUs). These devices determine the position of the mobile platform and the direction and orientation of the sensor at every moment [4].

Given that MLSs and ALSs capture data in large areas within short periods, both sensors are commonly used for urban applications, while the TLS is reserved for short-range applications, such as forest inventory [5], deformation monitoring [6] or heritage documentation [7]. ALS and MLS sensors provide three-dimensional (3D) point cloud data from mobile platforms, but significant differences exist between the two systems. ALS capture objects from the top view, while MLS and TLS collect data from the side-view, which makes the data taken by both types of sensors complementary. Additionally, the distance between the sensors and the ground is shorter in an MLS than in an ALS; consequently, the former performs measurements with higher resolution and greater density than the aerial sensors. ALS sensors cover large areas cost-effectively and rapidly but fail to capture details of small urban targets. Thus, MLS sensors are suitable for ground-based object modeling and to detect and extract elements located at street level, hardly achievable tasks in low density ALS data [8]. The main disadvantage is that MLS output files are large and hard to manage, forcing the development of organizing, cataloging, and optimizing methodologies to reduce the computation time significantly.

Many works in which point clouds are involved incorporate a preprocessing step or develop techniques that facilitate the treatment of the datasets and reduce the processing times. In some cases, a voxelization is performed to divide the point cloud space in a 3D grid of small regular cubes called voxels (volume elements) whose resolution depends on the size of the grid cells [9,10]. On a different approach, the point cloud is decomposed into several two-dimensional vertical slices using the global positioning system (GPS) time as auxiliary information [11] or into horizontal sections, parallel and above the ground [12]. Other works analyze each scan line individually instead of considering the cloud as a whole [13,14]. Removal of parts of the cloud that belong to objects that are not the focus of the study [15] is another common technique. A segmentation procedure is also routinely used for point cloud handling. Segmentation is the process of grouping the points of the cloud into segments: points in the same region are given the same category and treated as a set [16]. Some segmentation techniques such as graph cut [17], region growing [18], and 3D connected components [19] are also applied to facilitate the handling of the point cloud.

Creating and updating the databases of vegetation elements and street furniture in urban environments is an important issue in 3D city modeling, city management, and urban planning. Some cities such as Melbourne, Australia, have created their own street objects database in order to improve the design, amenity, and quality of the public environment [20]. The creation of these inventories with field visits and photo interpretation of remote sensing data can be an expensive, tedious, and imprecise work. Thus, recent studies have also started to address the automatic or semi-automatic computerized extraction of urban objects. Generally, these types of elements, whether trees, lampposts, or signs, are cylindrical or conical in geometry. In [21] and [22] two methods for detecting generic cylindrical elements using Hough transform and Random Sample Consensus (RANSAC), respectively, were proposed. In [23] and [24] the authors searched vertical isolated elements in a point cloud previously structured in voxels or regions within a previous segmentation procedure. [25,26] developed different methods that also depend on the geometry of vertical urban elements. In these cases, the detection is based upon the study of the three eigenvalues obtained from the covariance matrix of each segment in which the cloud was previously decomposed. In [27], trees were detected from a priori information of geometric features, such as the roughness and the point density ratio. Continuing this trend, in [28] a knowledge-based classifier that uses the size, shape, height, and reflectance intensity information of each pole as descriptors is proposed. Another useful technique consists of simplifying the 3D point cloud by projecting it in several 2D planes, both horizontal and vertical, and searching and classifying street objects represented in the cloud. This approach is followed in [29] where a method for extracting trees that voxelizes the point cloud and studies layers at different heights is developed. Potential trees are represented by the voxels that are isolated in consecutive layers. In [30] the authors segmented every scan line based on the distance between adjacent points; clusters were merged to group the segments that represented the same pole-like object. The classification between poles and non-poles was based on a priori information of geometric features such as the length of the cluster, its shape, direction, and number of sweeps. In [15] an algorithm for extracting lampposts was proposed in which a gridding process is applied to the point cloud. In every cell of the grid, the height of the highest point is stored; those cells that are taller than an established threshold are considered lampposts. An automatic method for extracting individual trees is presented in [31]. It consists of separating trees from man-made objects by projecting 3D points on horizontal grid accumulators at three heights and performing a cross comparison through these layers. In [32] the point cloud was projected in planes orthogonal to the direction of the MLS trajectory before the extraction of street curbs. Among the non-road segments, the street light poles were segmented using a pairwise 3D shape context based on a priori information of the type of lampposts of the area.

In our approach a fully automated method for detecting pole-like objects and classifying them as trees and man-made poles is developed. This method detects and classifies vertical urban elements from MLS data by means of a three step procedure:

1) A preprocessing stage, including a reference frame transformation and a region of interest (ROI) isolation. These procedures diminish the size of the original point cloud, the number of false positives in the following procedures, and the computational effort of the successive stages.

2) Vertical urban elements detection using the Reed and Xiaoli (RX) anomaly detection algorithm. Previously the preprocessed point cloud is organized in a pillar structure.

3) Vertical elements classification into two classes (trees and man-made poles) using an unsupervised classification algorithm.

2. METHOD

2.1. Preprocessing

Three-dimensional point cloud data files from MLS data systems include not only X, Y, and Z point coordinates but also additional information such as GPS time, scan angle, or reflectance intensity information, for the millions of points contained in the point cloud. In the current paper, the preprocessing step is divided into two main stages: (i) transforming the reference frame and (ii) removing the parts of the cloud that are not of interest in this work (point cloud reduction).

2.1.1. Reference Frame Transformation

Point clouds registered by MLS sensors are properly geo-referenced in a global reference system by a navigation system (GNSS) and an IMU, which provide coordinates within a global frame to every registered point. The original coordinate system is now transformed by means of a translation and three rotations into a local Cartesian coordinate system. The origin of the new reference frame is located at the beginning of the MLS trajectory, the z-axis is coincident with the local vertical direction, and the x-axis is coincident with the average direction of the vehicle. The y-axis completes the dextro-rotatory set, which makes local (x,y,z) coordinates handier than the global ones.

2.1.2. Removing Uninteresting Points

In an urban environment, objects such as columns of buildings, fences, or decorative elements on façades that are not of interest in this work can be found; they are not urban elements and, in addition, they can be wrongly detected as pole-like elements in the detection procedure. Normally, these elements are located in distant areas of the mobile laser scanner data, inside buildings or local businesses; meanwhile, vertical urban furniture and trees are located on the sidewalks and the surrounded area of the road. To avoid these false positives, a method to remove all these uninteresting points from the original point cloud was developed. The procedure consists of two steps: (i) an index is developed based on geometric features to determine the vertical (mainly buildings and fences) and large horizontal surfaces (roads and sidewalks); and (ii) the 3D connected components are segmented to group the points that are part of the same surface.

Geometric Index Definition

To identify the façades of the point cloud, a geometric index was developed. Indexes developed from geometric features of the cloud have been adopted in previous works such as [33], in which an operator based on a normal vector was introduced as a preprocessing step of an object recognition procedure. The index elaborated in the current paper is called the Geometric Index (GI) and combines the information provided by the normal vector and roughness values of every point of the cloud:

$$GI_i = \frac{|N_{Xi}| + |N_{Yi}| - |N_{Zi}|}{|N_{Xi}| + |N_{Yi}| + |N_{Zi}|} \times \frac{1}{e^{R_i}} \tag{1}$$

In Equation (1) (***Nxi***, ***NYi***, ***NZi***) are the components of the normal vector $\vec{N}$ in the point ***Pi*** and ***Ri*** is the roughness of the studied point. These values are measured from those points contained in a sphere of radius r centered in the studied point (***Pi***). Roughness (***Ri***) is defined as the distance between the studied point ***Pi*** and the least square best fitting plane comprising ***Pi*** and its neighborhood points inside the radius r sphere [6]. The first term of Equation (1) combines the three elements of the normal vector in a single value normalized in [–1, +1]. The behavior of the normal vector and its sensitivity to variations in the neighborhood size have been analyzed in five urban element types, easily identified in urban environments: façades, treetops, poles, roads, and cars. Significant differences have been found between these elements. In those elements with a horizontal flat surface that are determined as a trend surface (mainly roads and sidewalks), the vertical normal component (***NZi***) takes higher values than the horizontal ones ***NXi*** and ***NYi***. The opposite occurs on façades and fences, which are best fitted by vertical surfaces, in which horizontal normal vector components (***NXi*** and ***NYi***) take greater values than the vertical one. Other elements such as trees or cars present an irregular appearance because of their irregular and heterogeneous shapes. The roughness is included in the denominator of the second term of the GI Equation as an exponential to improve the separation between flat and rough surfaces. The lowest roughness values correspond to flat surfaces while higher roughness values take place in those elements with irregular shapes. According to the roughness study shown in Figure 1a, roughness ***Ri*** takes values around 0m in flat elements and higher values in rough surfaces; thus, term **1*eRi*** takes values around one for flat surfaces and lower in points that are further from the fitted plane. Consequently, the second term will not significantly affect the value of the GI in flat elements but will notably reduce it in rough surfaces, which helps to identify these elements in the point cloud. Since ***eRi*** is close to one for flat surfaces, this term has no effect on the GI but in contrast tends to substantially decrease the GI for rough surfaces, when **1*eRi*** is close to zero. To get the most suitable neighborhood size for the GI computation, normal and roughness features were studied in different radii values at the test sites. For small neighborhood radii (less than 20 cm), in many positions there are not enough point neighbors to compute the roughness and normal vector, making the distinction unclear. With large radii (more than 150 cm), the behavior of the GI in horizontal surfaces and elements at the ground level such as cars, pedestrians, or containers was quite similar. The neighborhood must be small enough not to consider points that belong to other elements but large enough to hold sufficient points to accommodate the interest features. Furthermore, the computational time increases exponentially with the radii and makes the process notoriously slow. Radii of 50 cm were set as optimal for extracting surfaces, because with this size (i) the GI values of flat surfaces, both horizontal and vertical, are suitably separated from other urban elements and (ii) the processing time is acceptable. In Figure 1, the roughness and the GI with different neighborhood sizes are shown. These studies were conducted in the two test sites. Figure 1a shows that the lowest roughness values correspond to flat surfaces; while higher roughness values correspond to the elements with irregular shapes. According to Figure 1b, the highest GI values, close to one, correspond to building façades and the lowest, around –1, to surfaces such as roads or pavement. Figure 2a shows the GI of the point cloud used as test site B in a color palette in which red corresponds to higher GI values, close to one, blue is reserved for the lowest GI values, and yellow and green represent the points with an intermediate GI value, around zero.

Extraction of Vertical and Horizontal Surfaces

To extract the vertical and horizontal surfaces, two thresholds αV and αH are set on the GI index. Those points Pi) with a higher GIi than αV are considered to belong to a vertical surface; meanwhile, points with a GIi below αH are treated as horizontal surfaces. Point clouds obtained after thresholding are composed of vertical and horizontal surfaces but also by points that satisfy these conditions that do not belong to these surfaces. These points are usually isolated or belong to small urban elements, such as treetops or pole-like objects.

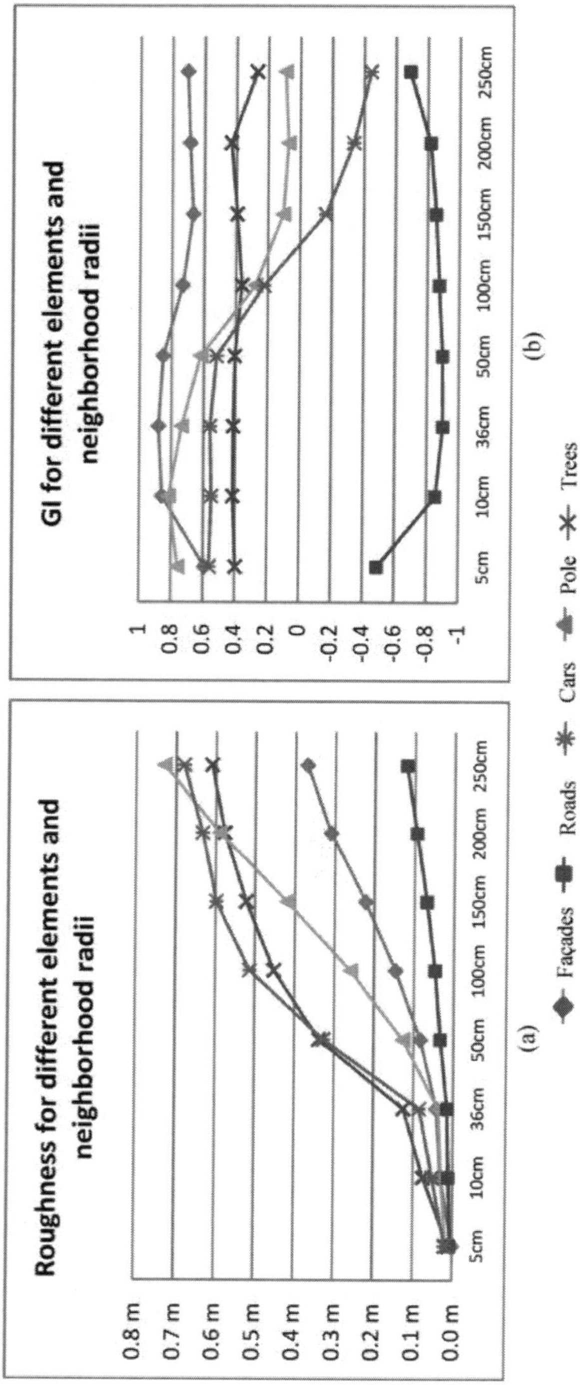

Figure 1. Behavior of the roughness (**a**) and GI (**b**) in five street elements for different neighborhood radii.

The 3D-connected components were segmented in favor of (i) grouping the points that belong to the same surface and (ii) removing isolated noisy points. Connected components analysis scans an image and labels its pixels into components if they are connected to each other (either four or eight connected) [34]. Once all groups have been determined, each pixel is labeled with an identifier according to the component the pixel was assigned to [35]. This technique is adapted to 3D point clouds structured in octrees. In a similar manner as for 2D images, the 3D connected components analyze the connectivity of the octrees and group in the same segment those that have a common side. In this case, the 3D connected components segmentation is defined by two parameters: the octree level (OL) and the minimum number of points per segment (MINP). The OL is related to the size of the octrees in which the point cloud is organized. It must be large enough for every octree not to be empty of points but sufficiently small for different urban elements to belong to independent octrees. A priori knowledge of the point cloud density is required to set the appropriate OL. Optimal OL has been empirically established, by the authors, as five times the mean distance between the points of the cloud. The MINP determines the number of components and their size. The objective of this step is removing isolated noisy points, and only large segments that represent building façades and pavement are considered. Once the 3D connected components are segmented, the entire segments recognized as façades are grouped into a single point cloud. This operation is repeated for the segments that represent roads, resulting in three point clouds: the original measured by the MLS sensor, one containing points that belong to building façades, and one with roads and sidewalks information (Figure 2b).

Original Point Cloud Reduction

The isolated region coming from preceding procedures that represents the road is analyzed using two-meter-wide sections, perpendicular to the x-axis of the local reference frame (Figure 3a). For each section, the center of the road and the location of building façades at both sides of the street are determined by analyzing the histogram of these point clouds. In every section, the road center is considered the modal class value in the y-coordinate histogram of the horizontal surfaces point cloud (Figure 3b). Thus, it is possible to approximately recreate the path followed by the MLS sensor, providing a kind of virtual MLS trajectory by joining the pavement center detected in each section. Additionally, for every 2-m-wide section, the alignment of the existing buildings is established by searching the modal class values of the y-coordinate histogram at both sides of the road center. A new point cloud is then generated by removing the points that lie beyond the façade line at both sides of the street (Figure 3d). This procedure automatically reduces the volume of the original point cloud, speeding up the following processes and removing potential false positives caused by vertical building columns. Furthermore, since this method is applied in narrow sections 2 m wide, it also accurately and precisely eliminates building façades in curved street sections or difficult areas, such as road intersections.

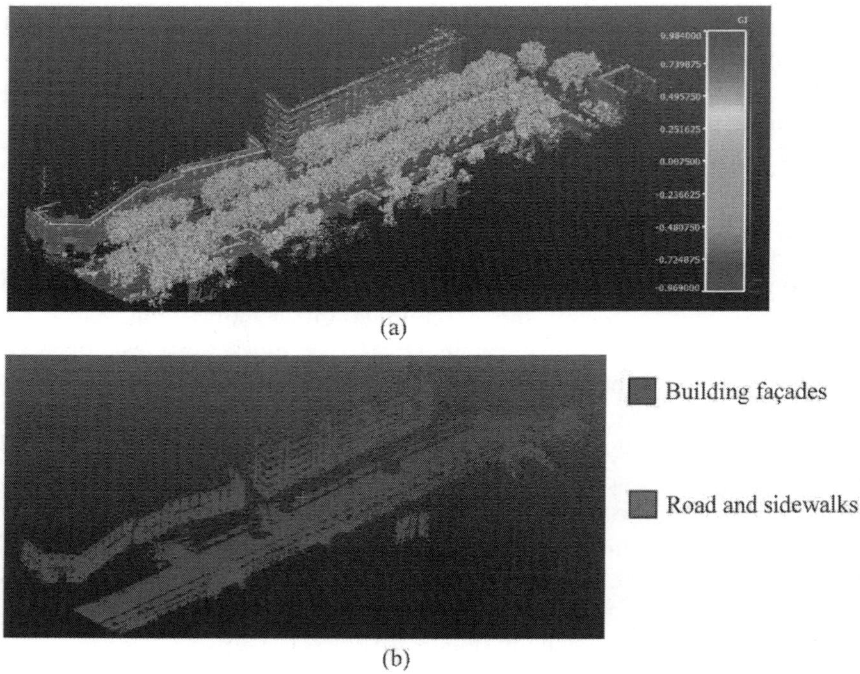

Figure 2. (**a**) GI in test site 2 and (**b**) façades detected after the connected components segmentation.

2.2. Pole-Like Elements Detection

2.2.1. Point Cloud Structuring

MLS data is composed of several million points so analyzing every single element and its neighborhood is computationally expensive and unproductive in terms of feature extraction. To speed up the detection and extraction procedure, the point cloud obtained in the previous step is organized and analyzed in a 3D vertical pillar structure pattern (Figure 4) [36].

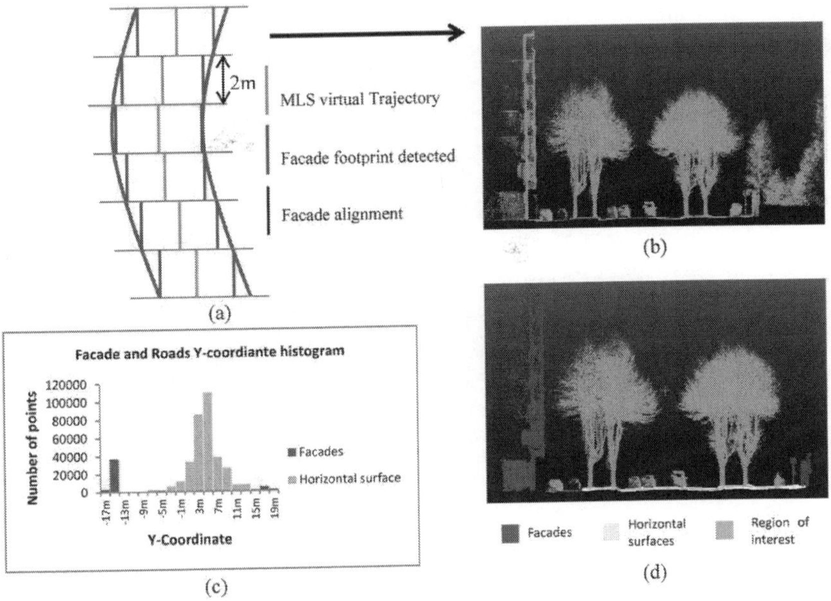

Figure 3. (**a**) Cloud MLS analysis in two meters width sections; (**b**) Original point cloud; (**c**) Histogram of façades and horizontal surfaces extracted and (**d**) Point cloud reduced: Isolated region of interest in green and removed façades in red.

Every point of the cloud is associated with a pillar, and all the points belonging to the same pillar are considered a set. The point cloud is divided in a 2D (m × n) grid composed of m columns and n rows. Each cell of the grid represents a pillar. Every pillar has a unique identifier ID assigned from its (x,y) coordinates in the 2D grid. Thus, for every cell $c_i = (x_i, y_i)$, with $x_i \in [0, m]$ and $y_i \in [0, n]$ corresponds the identifier $ID_c = (m \times y_i + x_i)$.

To avoid considering pillars as infinitely tall elements, the point distribution in each pillar is analyzed. This is achieved by decomposing the pillars into voxels of regular heights. The process starts searching the lowest occupied voxel, that is, the voxel with the lowest height that contains at least one point, and continues studying the voxels above it until an empty voxel is detected. Once a discontinuity is observed, that is, the first empty voxel above the occupied ones, the points above the discontinuity, if any, are discarded and not considered in the following steps. Thus, every pillar is formed by the points whose z-coordinate is between the lowest occupied voxel and the first discontinuity (empty voxel), found in the pillar. After this operation, every pillar is formed only by the elements connected to the ground level and disconnected points that unnecessarily increase the weight of the pillar and may hinder the detection and classification process are removed (Figure 5).

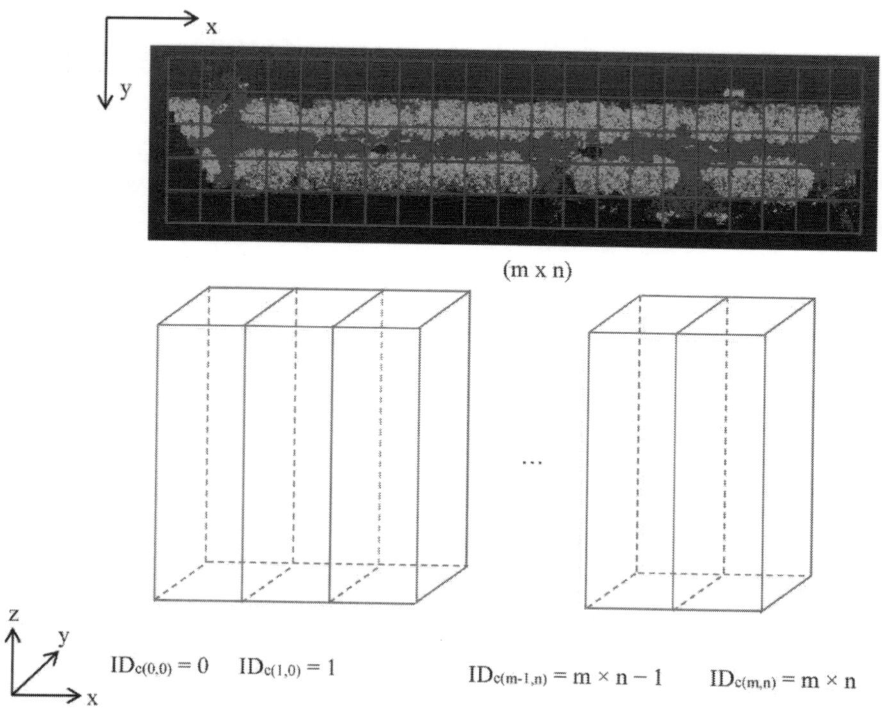

Figure 4. Creation of the pillar structure in the studied point cloud.

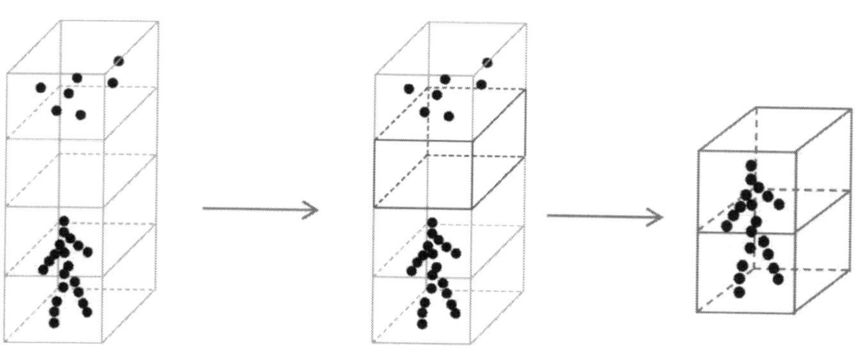

Figure 5. Pillar height is delimited until an empty voxel is found.

2.2.2. RX Anomaly Detection Algorithm

Once the point cloud has been structured, vertical urban elements are extracted and classified from the pillars in which the point cloud has been decomposed. It is necessary to determine which pillars contain a target element and which not. The RX anomaly detection algorithm is applied with this goal. This algorithm is commonly used to detect outliers in hyperspectral images, but it can also be used in multispectral images. The RX algorithm was developed by Reed and Xiaoli Yu [37]. It is based on the Mahalanobis distance and follows Equation (2) [38]:

$$\delta_{RX}(P_i) = (r_i - \mu)^T K_{LXL}^{-1}(r_i - \mu) \quad (2)$$

where ri is a vector in which considered features in the studied pillar Pi are saved, μ stores the mean values of the considered variables in the set of pillars of the whole point cloud, K is its sample covariance matrix, and L is the number of considered variables. i is the number of pillars in which the point cloud is structured. The minimum value of i is zero (the first studied pillar) and the maximum value depends on the size of the considered pillars The Mahalanobis distance is used to calculate how far each pillar is from the center of the cloud formed by the other pillars, and the shape of the cloud is considered through K. Mathematically, the RX algorithm performs some kind of inverse procedure of the principal component analysis (PCA); this was proved by Alonso et al. [39]. Anomalies should be understood as those elements whose spectral signature differs from the terrain in which they are. Anomalies are significant features of special interest to image analysts. In a hyperspectral image, every band contains information from a certain wavelength of the electromagnetic spectrum. The RX algorithm detects those pixels for which, in any band of the hyperspectral image, exists an anomalous spectral response compared with the response of the rest of the pixels of the image. MLS point clouds do not provide spectral information, but some geometric features can be computed for each point and its neighborhood. These geometric features have singular behaviors in vertical elements, quite different from other street elements.

In this work, the RX algorithm is applied to three features for every pillar of the point cloud. Height difference and the points' spatial dispersion have been considered to detect those pillars that represent a vertical urban element. To study the behavior of street objects in the variables, pillars that represent horizontal surfaces and vertical elements were chosen as ground truth (Table 1).

Height difference (Δh): every pillar is formed by points whose z-coordinate is between the lowest occupied voxel and the first voxel discontinuity. The height difference is referred to the distance, in terms of the z-coordinate, between the lowest and highest points of all points belonging to a pillar. The pillars in roads or pavement areas present a low height difference; however, vertical elements show a larger height difference between their lowest and highest points. Most pillars of an urban point cloud correspond to horizontal elements because they are the most common ones in streets environments. As can be seen in Table 1, the average value for the height difference in the full set of pillars is close to the trend of horizontal elements, with a low height difference (around 0.15 m).

Spatial dispersion is calculated from x- and y-coordinate dispersion (σ_X, σ_Y). The distribution of the (x,y) coordinates of the points contained in every pillar depends on which element is contained in it. The standard deviation of both planimetric coordinates (x,y) are the dispersion measures used as a geometric feature. In Table 1, the average (x,y) dispersion in roads and pavement is around 0.14 m; in pole-like elements, the average dispersion is a bit lower, around 0.10 m.

The number of points contained in every pillar (density) in which the point cloud is organized has been used in other works as a feature for extracting urban objects with satisfactory results [40]. Furthermore, surfaces that are orthogonal to the laser pulses show a higher density than those that are nearly parallel [11], a useful property for differentiating orthogonal from parallel elements. However, in the current work the accumulative number of points in every pillar was discarded and not included as a feature for detecting vertical elements. This is because the number of points that represent an urban element depends on the relative position of every element in relation to the MLS sensor and on the laser scanner properties. The same urban furniture located at both sides of the street does not have the same number of points in the 3D dataset even though they correspond to the same type of element. The closer an element is to the sensor, the more points represent it in the point cloud. Incorporating the point accumulation as a descriptor in an automatic detection procedure may cause errors in the process due to the different behavior of the elements shown in the point cloud.

To determine the relationship between the RX values and the features, the correlation between these variables was studied (Table 2). The RX values and height differences had a high positive correlation (0.72); meanwhile, the RX and both dispersions presented a negative correlation (–0.44 for $\boldsymbol{\sigma X}$ and –0.57 for $\boldsymbol{\sigma Y}$). The pillars with a ΔH higher than the average and dispersions ($\boldsymbol{\sigma X}$ and $\boldsymbol{\sigma Y}$) lower than the average have higher RX values. In Table 1, the mean value of the features (ΔH, $\boldsymbol{\sigma X}$, $\boldsymbol{\sigma Y}$) in three RX percentiles (P90, P95, and P99) are shown. As the correlation study suggested, the higher the RX values, the higher the ΔH and the lower the $\boldsymbol{\sigma X}$ and $\boldsymbol{\sigma Y}$. The pillars included in the RX 99th percentile are considered vertical urban elements since they perform a behavior similar to that of vertical elements' ground truth.

Table 1. Behavior of horizontal surfaces and pole-like elements in the considered features and average values among the full set of pillars and in percentiles 90, 95, and 99 of RX values.

	ΔH (m)	σ_x (m)	σ_y (m)
Ground Truth Horizontal surfaces	0.044	0.141	0.140
Ground Truth Vertical elements	2.78	0.103	0.098
Full pillars average	0.151	0.136	0.134
P_{90}	1.16	0.129	0.114
P_{95}	1.64	0.121	0.103
P_{99}	2.96	0.106	0.102

Table 2. Correlation matrix between RX and considered geometric features.

	ΔH	σ_x	σ_y
RX	0.72	−0.44	−0.57

2.3. Pole-Like Elements Classification

Once the vertical pole-like elements are extracted they are classified into two categories: man-made poles and trees. In this step each detected vertical element is isolated from the rest and treated as an independent set of points. The correct selection of descriptors is a key point to obtain good results in the classification procedure. In our case three descriptors for vertical element were computed: the roughness of their points (both mean and dispersion values) and the scattering of radial distance (ρ) of the cylindrical coordinate frame centered in the studied pole-like set of points.

Cylindrical coordinates: After the reference frame transformation performed in the preprocessing step, the point cloud is referred to a local coordinate system. In the current step a new reference frame transformation is performed for every detected pole, moving from the Cartesian local reference frame (x,y,z) to a cylindrical coordinate system (ρ, ϕ, z). For every detected pole-like object its own cylindrical coordinate frame system is established. Its cylindrical axis coincides with the direction of Z-axis in the local coordinate system and it is located in the (x,y) centroid of the set of the points that belong to the pole-like object. From the cylindrical triplet of coordinates, the most interesting feature to accomplish this classification is the radial distance (ρP). This is because points that belong to man-made poles are closely located around the vertical cylindrical axis than those that represent trees due to their thin appearance. Thus, the dispersion of $\boldsymbol{\rho P}$ in these elements is lower than in trees.

Roughness: It has been observed that both mean and standard deviation of roughness have a different behavior in each category, being their values significantly differ in both types of pole objects. Roughness values of artificial poles are lower than trees due to their flat and smooth shape on their upper part, contrary to the irregular and rough appearance of treetops, which cause higher values on these descriptors. Additionally, dispersion of this parameter in poles is lower than in trees due to the heterogeneity caused by branches and treetops

In order to test whether the geometric descriptors taken into account are distinguishable and present a distinctive behavior in the two considered classes, a separability study has been carried out. To achieve this inspection a ground truth has been generated by identifying diverse elements of both categories in the point cloud. There are several methods to measure the separability between variables; in this work Jeffries-Matusita (JM) distance and transformed divergence, computed from Bhattacharyya distance (BD) (3) has been used as separability measure [41]. In Equation (3), ($\mu\boldsymbol{a}$,$\mu\boldsymbol{b}$) and ($\boldsymbol{\sigma a}$,$\boldsymbol{\sigma b}$) are, respectively, the mean and standard deviation of classes a and b. JM distance (4) takes values in the range [0,2]. The higher JM values, the higher the separability between the studied classes. As can be seen in Table 3, differences between man-made poles and trees are considerably higher in the three examined variables, taken values around 1.5 and 1.8 for mean and dispersion roughness respectively, and above 1.5 in the radial distance. According with the given separability values, it is expected to obtain accurate results by the clustering algorithm in the classification of man-made poles and trees.

$$BD = \frac{1}{8} \times (\mu_a - \mu_b)^2 \times \frac{2}{(\sigma_a^2 + \sigma_b^2)} + \frac{1}{2} \times \log(\frac{\sigma_a^2 + \sigma_b^2}{2 \times \sigma_a \times \sigma_b}) \quad (3)$$

$$JM = 2 \times (1 - e^{-BD}) \quad (4)$$

Table 3. Jeffries-Matusita distances for man-made poles and trees in the considered descriptors: mean and standard deviation of roughness (μ_r and σr) and standard deviation of ρ ($\sigma\rho$).

Test Case	μ_r poles-trees	σ_r poles-trees	σ_ρ poles-trees
Dataset A	1.51	1.76	1.72
Dataset B	1.52	1.88	1.53

3. TEST CASES

The efficiency of the proposed method was tested in two datasets measured by different MLS sensors. In every test site the detection and classification procedure have been performed in order to test the capability of the proposed method to extract and classify pole-like objects.

3.1. Mapping Data

3.1.1. Dataset A

The point cloud used as test site 1 represents a 300 m section of an urban street in Boadilla del Monte, a city in western Madrid, Spain. This street is a type of a wide boulevard, with two lanes for each direction, the tracks of a tram in the median strip, and sidewalks and parking areas on both sides of the street. Features such as trees, shrubbery, traffic lights, lampposts, containers, bus shelters, pedestrians, or vehicles are present in this scene. This dataset was selected to test the method in areas of the city with wide streets and a great variety of vertical elements. The slope, 5% on average in this street section, also affected the selection of this test site. This dataset comprises more than 3 million points and was acquired with the IP-S2 Compact + system produced by Topcon Inc. The IP-S2 incorporates a dual frequency GNSS receiver, an IMU, and a connection to external wheel encoders, which receive odometry information. These three systems provide a highly-accurate 3D position for the vehicle. The IP-S2 Compact + scanner is equipped with five laser scanners that collect 150,000 points per second at a range of 40 m, with a vertical field of view of 360°. It is also equipped with a panoramic camera that delivers 360° spherical imagery.

3.1.2. Dataset B

A dataset corresponding to test site 2 was measured by a Lynx Mobile Mapper system, produced by Optech Inc. The Lynx scanner collects survey-grade LIDAR data at 500,000 measurements per second with a 360° field of view (FOV). The Lynx also incorporates the POS LV 520, by Applanix, which integrates an IMU with a two-antenna heading measurement system. LIDAR sensors are located in the rear of a van. Each sensor registers points in a plane at 60° to the horizontal and 45° to the longitudinal axis of the driving direction. This laser scanner provides absolute accuracies of 0.015° in heading, 0.005° in roll and pitch, 0.02 m in the X, Y positions, and 0.05 m in the Z position. All values are determined via differential GPS post-processing after data collection using GPS base station data [42]. In this case, the point cloud was composed of more than 6 million points, and the measurements were made along a 400-m-long street in Busto Arsizio, in the Lombardy region, in northern Italy. The street is narrow, and there is one lane in each direction and sidewalks, parking areas, and buildings on both sides of the road. Furthermore, there is a double barrier of leafy tall trees on both sides of the road that causes occlusions in urban furniture, such as lampposts or traffic signs present in this test site. This site was chosen to test the efficiency of the method in narrow streets covered by dense woody vegetation.

3.2. Reference Data

A ground truth was created in each of the two datasets in order to evaluate the results provided by the detection and classification procedures. The target elements included in the detection ground truth database are those with a pole-like shape, among which are lampposts, traffic signs, traffic lights, and trees. In the classification reference data, pole-like objects are sorted into two categories: man-made poles and trees. The reference datasets were composed of all the pole-like elements that were identifiable in the original point cloud. Ground truth in Dataset A is composed of 241 pole-like objects; 141 were man-made poles and 100 were trees. In Dataset B, a total of 228 pole-like elements were observed; 56 were trees and 172 artificial poles.

The validity of our model was quantified by means of completeness, correctness and quality quantifiers, which follow Equations (5)–(7), respectively [43]. TP (true positive) are the detected poles that matched the reference, FP (false positive) represents the detected poles that do not match the ground truth, and FN (false negative) symbolize the poles that exist in the ground truth but are not detected by the proposed method.

$$Completeness = \frac{poles\ matched\ the\ reference}{poles\ of\ reference} = \frac{TP}{TP + FN} \quad (5)$$

$$Correctness = \frac{poles\ matched\ reference}{extracted\ poles} = \frac{TP}{TP + FP} \quad (6)$$

$$Quality = \frac{poles\ matched\ the\ reference}{extracted\ poles\ +\ unmatched\ reference} = \frac{TP}{TP + FP + FN} \quad (7)$$

To quantify the results of the classification step, the classification ground truth was compared with the labeled point cloud provided by the clustering algorithm. For every test site, a confusion matrix was constructed from which five parameters well-known in the evaluation of classification procedures are extracted: overall accuracy, commission and omission errors, and user and producer accuracy [44].

3.3. Algorithm Settings

One of the main purposes of this work was to develop automated extraction and classification procedures, which minimize user interaction. To achieve this goal, the variables and the parameters must be robust enough to be independent of the attributes of the point cloud and the configuration of the study street. We determined that the critical parameters are (i) thresholds in the geometric index to extract vertical and horizontal surfaces, (ii) the settings of the connected components segmentation, and (iii) the percentile of RX values that represent pole-like objects. The sensibility of each parameter has been analyzed in order to establish the range of values that every parameter can take without affecting the final result of the procedure (Table 4). Regarding RX percentile, which is the parameter that determines the pole-like objects detection, its influence in the extraction has been studied and quantified for different percentile values in order to determine the optimal ones. It has been concluded that RX percentile values that provide the best quality rates are $P_{98.5}$ and P_{99} (Table 5).

In the current work, the GI thresholds were set from the studies summarized in Figure 1, in which the vertical surfaces (façades) were detected for $\alpha_V >$ 0.8 and horizontal elements (pavement and sidewalks) were located when α_H > -0.8. Thus, the vertical and horizontal surfaces were set to $GI_i > 0.8$ and $GI_i < -0.8$, respectively. Regarding the 3D connected components segmentation, the MINP was set to 2000 points/region. For the octree level (OL), in the cases the mean distance between points was almost 4 cm, which implies an OL of 20 cm. Other parameters, such as pillar size and RX, are less dependent on the characteristics of the cloud and had similar values in every case because they refer to the properties of pole-like urban elements. For the test sites used in this work, the pillar size was established at 50×50 cm, and the RX percentile was fixed at P_{99}. The same settings were applied to both test sites (Table 4).

Table 4. Algorithm settings used in the test sites A and B and range of values that every parameter can take.

	Algorithm Settings	Parameter Ranges
Vertical surfaces threshold (α_V)	0.8	[0.75; 0.85]
Horizontal surfaces threshold (α_H)	−0.8	[−0.9; −0.7]
Minimum number of points (MINP)	2000 points	[2000; 3000]
Octree level (OL)	20 cm	[15cm ; 25cm]
Pillar size	0.5m	[0.4m; 0.6m]
RX Percentile	P_{99}	[$P_{98.5}$; P_{99}]

4. RESULTS

4.1. Dataset A

In the point cloud corresponding to this dataset, 241 pole-like elements have been observed among trees, lampposts, traffic signs, traffic posts, and tram posts. The detection procedure extracts 233 vertical elements (Figure 6a), of which 230 match with the ground truth reference and the three remaining detected poles correspond with a working vehicle that has a similar structure to the pole-like objects (Figure 7e). Consequently, eleven poles were undetected, nine of them due to their position in occluded or shadowed regions of the point cloud (Figure 7c). The two others non-detected poles are traffic signs of low height included in the reference dataset, but not high enough to be extracted by the method (Figure 7a,b). According with these results, the detection step takes completeness, correctness, and quality rates of 95.4%, 98.7%, and 94.3%, respectively (Table 5).

Regarding the classification procedure, 217 out of 230 vertical elements were correctly labeled (Figure 6b), which means an overall accuracy of 94.35% (Table 6). About trees category, 85 trees were correctly labeled and eight were wrongly classified as artificial poles due to their scarce and sparse vegetation, similar to that of a man-made pole (Figure 7i). In terms of producer's and user's accuracy, 91.4% and 94.44% are obtained in trees category with commission and omission errors of 5.56% and 8.6%, respectively. In relation to poles, five of 137 man-made poles included in the ground truth reference were incorrectly labeled. These poles are close to trees and their branches modify the appearance of the artificial poles, providing a scattered shape more characteristic of trees than of its own nature (Figure 7f,g). This classification results in a commission and omission errors of 5.71% and 3.65%, being the producer’s and user’s accuracy achieved 96.35% and 94.29%, respectively (Table 6). In an overall evaluation of detection and classification procedures, 217 pole-like objects out of 241 were correctly detected and classified, which means an accuracy of 90.04% (Table 7).

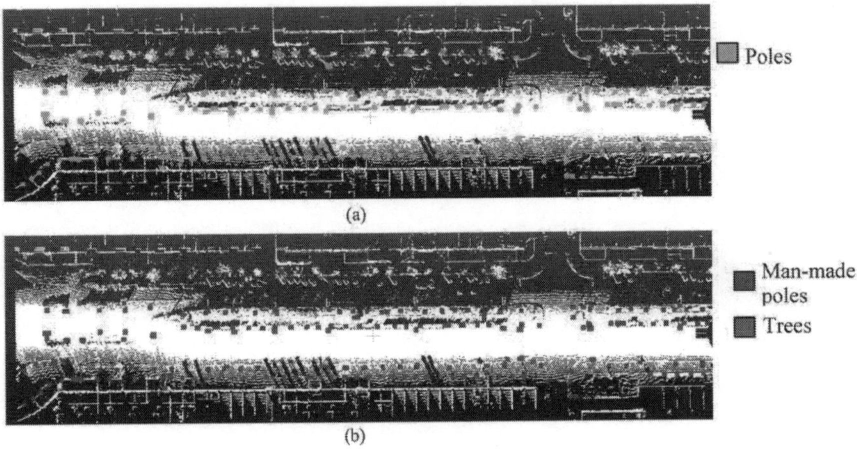

Figure 6. Results for the detection (**a**) and classification (**b**) procedure in Test Site 2.

Table 5. Completeness, correctness and quality achieved with the proposed detection method in the two studied test sites with different RX percentile values.

Test Site A/ RX Percentile	Observed	Detected	FP	FN	TP	Completeness	Correctness	Quality
97.5	241	347	111	5	236	97.93	68.01	67.05
98	241	314	78	5	236	97.93	75.16	73.98
98.5	241	252	17	6	235	97.51	93.25	91.09
99	241	233	3	11	230	95.4	98.7	94.3
99.5	241	144	2	99	142	58.92	98.61	58.44
Test Site B/ RX Percentile	**Observed**	**Detected**	**FP**	**FN**	**TP**	**Completeness**	**Correctness**	**Quality**
97.5	228	359	136	5	223	97.81	60.43	61.26
98	228	314	91	5	223	97.81	68.83	69.91
98.5	228	244	21	5	223	97.81	91.39	89.56
99	228	222	2	8	220	96.5	99.1	95.7
99.5	228	119	1	110	118	51.75	99.16	51.53

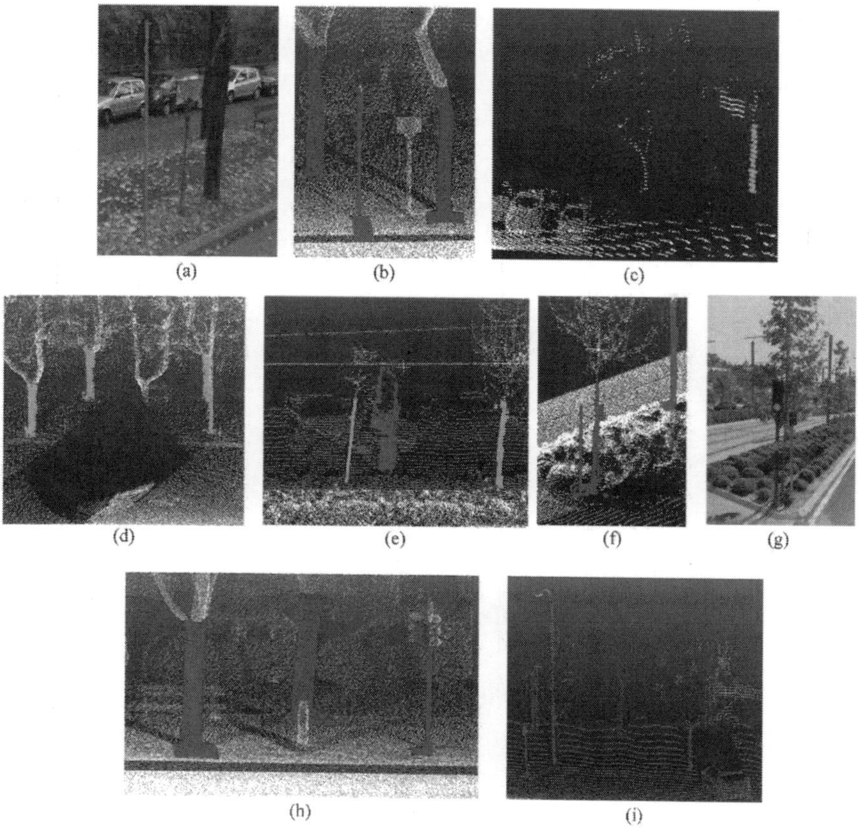

Figure 7. (**a,b**) little signals appearance in a RGB image and in the point cloud, (**c,d**) occlusion of a tree trunk, (**e**) in red, features detected in a work vehicle, (**f,g**) man-made pole surrounded by tree branches in a RGB image and in the point cloud, (**h**) rough and scattered man-made traffic light wrongly classified as a tree, and (**i**) tree with scarce and sparse vegetation misclassified as a man-made pole.

4.2. Dataset B

In this street section, 228 vertical elements were observed of which 220 were correctly extracted, eight were undetected, and two were falsely detected. Thus, the completeness of the detection procedure was higher than 96%, the correctness above 99%, and the quality higher than 95% (Figure 8a,b and Table 5). Regarding the eight false negatives, the undetected elements, seven were discarded by the method because they were a special kind of traffic sign, with a lower height than ordinary signs (Figure 7a,b). The remaining missing pole element corresponded to a tree trunk, which was partially occluded in the point cloud by a parked van (Figure 7d). The two false positives were detected from the structure of a van that had a shape similar to pole-like objects, with high height differences and low dispersion in (x,y) coordinates. The number of non-target pole-like elements detected would have increased, especially inside the footprint of buildings, unless the original point cloud reduction step had not been carried out. In relation to the classification step, in this road section the overall accuracy rate was 95.0%, which means that 209 out of 220 vertical elements were correctly labeled (Figure 8c,d). According to Table 8, ten man-made poles were mistakenly labeled as trees. Six of these poles were surrounded by branches of nearby trees, which caused the scattered and rough appearance of their pole in the cloud. The remaining four poles were low traffic lights, which were misclassified due to the roughness generated by their upper light structure (Figure 7h). Only one tree was wrongly classified as an artificial pole. The shape of this tree was similar to a pole, with a thin, tall trunk and barely scattered branches. These results provide a commission and omission rate of 6.02% and 0.63% in the trees and 1.85% and 15.87%, respectively, in the man-made poles category. For accuracy, tree classification achieved a producer accuracy of 99.36% and a user accuracy of 93.98%; meanwhile, the pole labeling was 84.13% and 98.15% in the producer and user accuracy, respectively (Table 8). Thus, 209 vertical elements out of 228 were correctly detected and labeled, which meant a global accuracy of 91.67% of the complete procedure (Table 7).

Table 6. Confusion matrix of the classification procedure in test site A, where columns are the ground truth and rows represent the classification results.

Dataset A	Trees	Poles	Σ
Trees	85	5	90
Poles	8	132	140
Σ	93	137	230
Overall Accuracy = 94.35% (217/230)			

Dataset A	Commission	Omission
Trees	5.56% (5/90)	8.6% (8/93)
Poles	5.71% (8/140)	3.65% (5/137)
	Producer's Accuracy	**User's Accuracy**
Trees	91.4% (85/93)	94.44% (85/90)
Poles	96.35% (132/137)	94.29% (132/140)

Table 7. Quality evaluation complete procedure in Dataset A and B.

	Observed	Detected	FP	Undetected	Correctly Labeled	Wrongly Labeled	Accuracy
Test site A	241	233	3	11	217	13	90.04 (217/241)
Test site B	228	222	2	8	209	11	91.67 (209/228)

Table 8. Confusion matrix of the classification procedure in Dataset B, where columns are the ground truth and rows represent the classification results.

Dataset B	Trees	Poles	Σ
Trees	156	10	166
Poles	1	53	54
Σ	157	63	220
	Overall Accuracy = 95.0% (209/220)		
Dataset B	**Commission**	**Omission**	
Trees	6.02% (10/166)	0.63% (1/157)	
Poles	1.85% (1/54)	15.87% (10/63)	
	Producer's Accuracy	**User's Accuracy**	
Trees	99.36% (156/157)	93.98% (156/166)	
Poles	84.13% (53/63)	98.15% (53/54)	

4.3. Comparison with Previous Methods

The results provided by our method were compared with other algorithms to evaluate its performance. In the current literature, several methods focus on extracting urban objects. In [15] the target elements were lampposts; trees were the main category in [27], and [23] extracted all types of pole-like objects without differentiating between different types. The lack of a common dataset with a ground truth associated means that every work uses its own dataset and creates a ground truth with visual inspections of the field and the point cloud. In [23] a method for extracting pole-like objects is presented that achieves a completeness detection rate average of 92.3% and a correctness of 83.8% in the four datasets. Most false positives obtained by this method are detected inside the footprint of buildings. The method developed by [30] achieved completeness and correctness rates of 77.7% and 81.0%, respectively, for targets closer than 30 m to the scanner route, which means a mean accuracy of 79.3%. Only when targets closer than 12.5

m were considered, these rates increased, achieving 83.5% and 86.5% for completeness and correctness, respectively, and a mean accuracy of 85%. Most failures in the remotest parts of the clouds were due to shadowed areas and low point density in these areas. [32] extracted lampposts in six datasets achieving completeness rates above 99% and correctness between 97.55% and 99.01; the quality index ranged from 96.74% to 98.21%. Despite the high accuracy, this method seems to be far from automated due to the large number of thresholds to be set to conduct the extraction. Individual street trees were extracted in [29] and accuracy rates above 98% were achieved. This method presents some limitations because it is designed to be used in flat terrains, and all trees must be the same height from the ground. Furthermore, this method achieves accurate results in individual street trees, but its effectiveness in dense vegetated areas where treetops are merged has not been tested. In [26], the accuracy in pole-like objects recognition was 63.9%, and in [31] the completeness and correctness achieved in detecting individual trees ranged from 80.7% to 81.2% and from 70.2% to 75.5%, respectively.

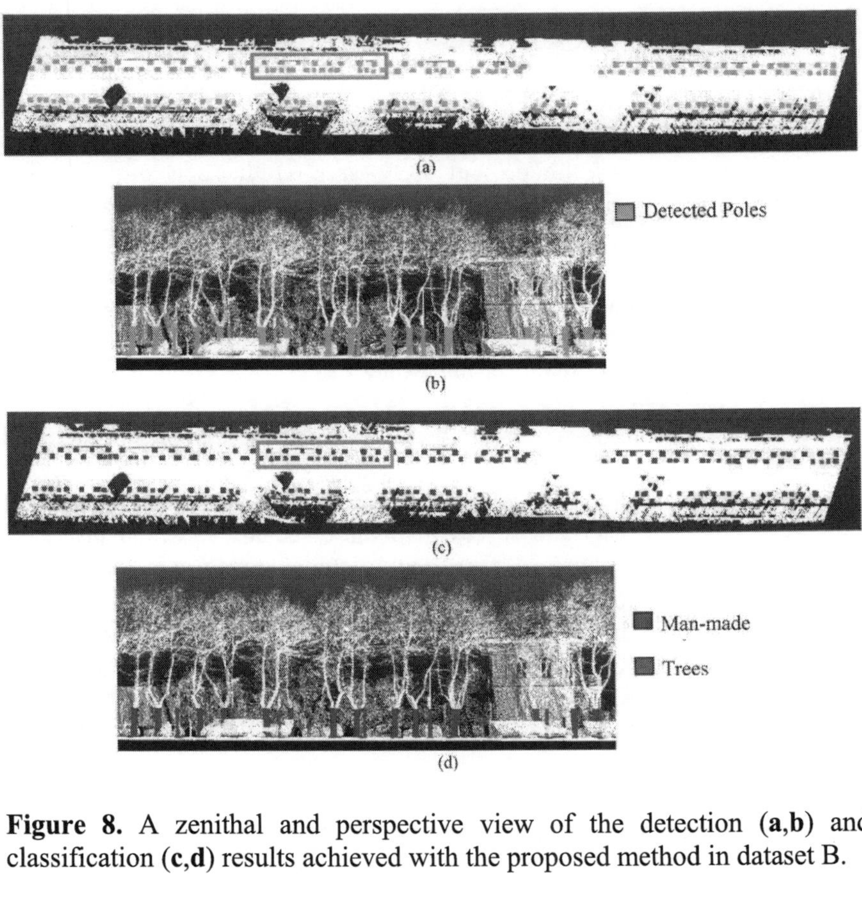

Figure 8. A zenithal and perspective view of the detection (**a**,**b**) and classification (**c**,**d**) results achieved with the proposed method in dataset B.

The pole-like object detection method proposed in this paper achieved quality rates in the two datasets of 94.3% and 95.7%, respectively, which are slightly higher than some of the previous methods. The two datasets used to test this method were measured by different sensors in diverse scenarios, which prove its robustness. This algorithm is independent of the scanning geometry and of the slope of the street because the coordinates are transformed in the preprocessing step. In addition, this process detects the horizontal and vertical surfaces in the point cloud and automatically delimits the regions of interest, thus avoiding false positives inside the footprints of buildings. Furthermore, this process does not require a priori information or previous training, and the number of thresholds has been minimized in order to automate the procedure. However, certain problems may occur in remote areas of the cloud with low point density and in trees whose trunks appear tilted. A previous work [30] proposed the development of methods for separating tree trunks from other poles. In the current work, trees and man-made poles were distinguished with a clustering algorithm. This classification procedure achieved an overall accuracy higher than 90% in every data case.

5. CONCLUSIONS AND FUTURE WORKS

The main novelties of the present work are: (i) the development of a geometric index that extracts horizontal and vertical surfaces and can also be used to reconstruct the MLS vehicle trajectory, (ii) the detection of pole-like objects by means of an anomaly detection algorithm and their classification in trees and man-made elements without prior training data, and (iii) the definition of a robust procedure that can be easily automated and provides accurate results with minimal user intervention.

The pole-like object detection method proposed in this paper achieved quality rates in the two datasets of 94.3% and 95.7%, respectively, which are slightly higher than other methods. The two datasets used to test this method were measured by different sensors in diverse scenarios, which prove its robustness. This algorithm is independent of the scanning geometry and of the slope of the street because the coordinates are transformed in the preprocessing step. In addition, this process detects the horizontal and vertical surfaces in the point cloud and automatically delimits the regions of interest, thus avoiding false positives inside the footprints of buildings. Furthermore, this process does not require a priori information or previous training, and the number of thresholds has been minimized in order to automate the procedure. However, certain problems may occur in remote areas of the cloud with low point density and in trees whose trunks appear tilted.

The typology and casuistry of urban pole-like objects are very diverse, and there is probably no single best method for detecting and classifying them in all cases. According to the results in this work, we can conclude that this method is robust, useful for automatically detecting and classifying pole-like objects, and provides satisfactory results regardless the heterogeneity of the area and the specifications of the sensor and does not need the knowledge of the measured MLS trajectory. In the future, other anomaly detection algorithms could be tested in the detection step, and other features such as laser intensity could be introduced in the classification procedure to expand the classification to other types of urban elements. The detecting procedure provided quality values of around 95% in two test sites, and the classification step achieved an overall accuracy above 94%. In an overall evaluation of both procedures, more than 90% of the vertical elements were correctly detected and classified.

ACKNOWLEDGMENTS

The authors would like to thank the Optech Inc. and TOPCON Inc. for providing the datasets to test the proposed method.

AUTHOR CONTRIBUTIONS

The authors, Borja Rodríguez-Cuenca, Silverio García-Cortés, Celestino Ordóñez and Maria C. Alonso, designed the research, performed data analysis, and contributed with ideas, writing and discussion.

CONFLICTS OF INTEREST

The authors declare no conflict of interest.

REFERENCES AND NOTES

1. Mundia, C.; Aniya, M. Analysis of land use/cover changes and urban expansion of Nairobi city using remote sensing and GIS. Int. J. Remote Sens. **2005**, 26, 2831–2849.

2. Dimitrakopoulos, G.; Demestichas, P. Intelligent transportation systems. IEEE Veh. Technol. Mag. **2010**, 5, 77–84.

3. Levinson, J.; Askeland, J.; Becker, J.; Dolson, J.; Held, D.; Kammel, S.; Kolter, J.Z.; Langer, D.; Pink, O.; Pratt, V. Towards fully autonomous driving: Systems and algorithms. In Proceedings of IEEE Intelligent Vehicles Symposium (IV), Baden-Baden, Germany, 5–9 June 2011; pp. 163–168.

4. Jiang, X.; Bunke, H. Fast segmentation of range images into planar regions by scan line grouping. Machine Vis. Appl. **1994**, 7, 115–122.

5. Hyyppä, J.; Jaakkola, A.; Chen, Y.; Kukko, A. Unconventional LiDAR mapping from air, terrestrial and mobile. In Proceedings of the Photogrammetric Week, Stuttgart, Germany, 9–13 September 2013; pp. 205–214.

6. Zogg, H.; Ingensand, H. Terrestrial laser scanning for deformation monitoring—Load tests on the Felsenau Viaduct (CH). Int. Arch. Photogramm. Remote Sens. **2008**, 37, 555–562.

7. Rüther, H.; Held, C.; Bhurtha, R.; Schröder, R.; Wessels, S. Challenges in heritage documentation with terrestrial laser scanning. In Proceedings of the 1st AfricaGEO Conference, Capetown, South Africa, 30 May–2 June 2011.

8. Zhu, L.; Hyyppa, J. The use of airborne and mobile laser scanning for modeling railway environments in 3D. Remote Sens. **2014**, 6, 3075–3100.

9. Vosselman, G.; Gorte, B.G.; Sithole, G.; Rabbani, T. Recognising structure in laser scanner point clouds. Int. Arch. Photogramm. Remote Sens. Spat. Inf. Sci. **2004**, 46, 33–38.

10. Douillard, B.; Underwood, J.; Kuntz, N.; Vlaskine, V.; Quadros, A.; Morton, P.; Frenkel, A. On the segmentation of 3D LiDAR point clouds. In Proceedings of IEEE International Conference on Robotics and Automation (ICRA), Shanghai, China, 9–13 May 2011; pp. 2798–2805.

11. Yang, B.; Fang, L.; Li, J. Semi-automated extraction and delineation of 3D roads of street scene from mobile laser scanning point clouds. ISPRS J. Photogramm. Remote Sens. **2013**, 79, 80–93.

12. Rodríguez-Cuenca, B.; García-Cortés, S.; Ordóñez, C.; Alonso, M.C. An approach to detect and delineate street curbs from MLS 3D point cloud data. Autom. Constr. **2015**, 51, 103–112.

13. Jaakkola, A.; Hyyppä, J.; Kukko, A.; Yu, X.; Kaartinen, H.; Lehtomäki, M.; Lin, Y. A low-cost multi-sensoral mobile mapping system and its feasibility for tree measurements. ISPRS J. Photogramm. Remote Sens. **2010**, 65, 514–522.

14. Lehtomäki, M.; Jaakkola, A.; Hyyppä, J.; Kukko, A.; Kaartinen, H. Performance analysis of a pole and tree trunk detection method for mobile laser scanning data. Int. Arch. Photogramm. Remote Sens. Spat. Inf. Sci. **2011**, 38, 197–202.

15. Yujie, H.; Xiang, L.; Jun, X.; Lei, G. A novel approach to extracting street lamps from vehicle-borne laser data. In Proceedings of 19th International Conference on Geoinformatics, Shanghai, China, 24–26 June 2011; pp. 1–6.

16. Rabbani, T.; van den Heuvel, F.; Vosselmann, G. Segmentation of point clouds using smoothness constraint. Int. Arch. Photogramm. Remote Sens. Spat. Inf. Sci. **2006**, 36, 248–253.

17. Golovinskiy, A.; Funkhouser, T. Min-cut based segmentation of point clouds. In Proceedings of IEEE 12th International Conference on Computer Vision Workshops (ICCV Workshops), Kyoto, Japan, 27 September–4 October 2009; pp. 39–46.

18. Belton, D.; Lichti, D.D. Classification and segmentation of terrestrial laser scanner point clouds using local variance information. In Proceedings of ISPRS Commission V Symposium: Image Engineering and Vision Metrology (IAPRS), Dresden, Germany, 25–27 September 2006.

19. Verma, V.; Kumar, R.; Hsu, S. 3D building detection and modeling from aerial LiDAR data. In Proceedings of IEEE Computer Society Conference on Computer Vision and Pattern Recognition, New York, USA, 17–22 June 2006; pp. 2213–2220.

20. City of Melbourne Street Furniture Plan. Available online: http://www.melbourne.vic.gov.au/ (accessed on 22 June 2015).

21. Rabbani, T.; Van Den Heuvel, F. Efficient hough transform for automatic detection of cylinders in point clouds. In Proceedings of ISPRS Workshop: Laser Scanning 2005, Enschede, The Netherlands, 12–14 September 2005; pp. 60–65.

22. Schnabel, R.; Wahl, R.; Klein, R. Efficient RANSAC for point-cloud shape detection. Comput. Graph. Forum **2007**, 26, 214–226.

23. Cabo, C.; Ordoñez, C.; García-Cortés, S.; Martínez, J. An algorithm for automatic detection of pole-like street furniture objects from mobile laser scanner point clouds. ISPRS J. Photogramm. Remote Sens. **2014**, 87, 47–56.

24. Brenner, C. Extraction of features from mobile laser scanning data for future driver assistance systems. In Advances in Giscience; Springer: Berlin/Heidelberg, Germany, 2009; pp. 25–42.

25. El-Halawany, S.I.; Lichti, D.D. Detection of road poles from mobile terrestrial laser scanner point cloud. In Proceedings of International Workshop on Multi-Platform/Multi-Sensor Remote Sensing and Mapping (M2RSM), Xiamen, China, 10–12 January 2011; pp. 1–6.

26. Yokoyama, H.; Date, H.; Kanai, S.; Takeda, H. Pole-like objects recognition from mobile laser scanning data using smoothing and principal component analysis. Int. Arch. Photogramm. Remote Sens. Spat. Inf. Sci. **2011**, 38, 115–120.

27. Rutzinger, M.; Pratihast, A.; Oude Elberink, S.; Vosselman, G. Detection and modelling of 3D trees from mobile laser scanning data. Int. Arch. Photogramm. Remote Sens. Spat. Inf. Sci. **2010**, 38, 520–525.

28. Li, D.; Elberink, S.O. Optimizing detection of road furniture (pole-like objects) in mobile laser scanner data. ISPRS Ann. Photogramm. Remote Sens. Spat. Inf. Sci. **2013**, 1, 163–168.

29. Wu, B.; Yu, B.; Yue, W.; Shu, S.; Tan, W.; Hu, C.; Huang, Y.; Wu, J.; Liu, H. A voxel-based method for automated identification and morphological parameters estimation of individual street trees from mobile laser scanning data. Remote Sens. **2013**, 5, 584–611.

30. Lehtomäki, M.; Jaakkola, A.; Hyyppä, J.; Kukko, A.; Kaartinen, H. Detection of vertical pole-like objects in a road environment using vehicle-based laser scanning data. Remote Sens. **2010**, 2, 641–664.

31. Yao, W.; Fan, H. Automated detection of 3D individual trees along urban road corridors by mobile laser scanning systems. In Proceedings

of International Symposium on Mobile Mapping Technology (MMT), Tainan City, Taiwan, 1–3 May 2013.

32. Yu, Y.; Li, J.; Guan, H.; Wang, C.; Yu, J. Semiautomated extraction of street light poles from mobile lidar point-clouds. IEEE Trans. Geosci. Remote Sens. **2015**, 53, 1374–1386.

33. Ioannou, Y.; Taati, B.; Harrap, R.; Greenspan, M. Difference of normals as a multi-scale operator in unorganized point clouds. In Proceedings of 2nd International Conference on 3D Imaging, Modeling, Processing, Visualization and Transmission (3DIMPVT), Zurich, Switzerland, 13–15 October 2012; pp. 501–508.

34. Di Stefano, L.; Bulgarelli, A. A simple and efficient connected components labeling algorithm. In Proceedings of International Conference on Image Analysis and Processing, Venice, Italy, 27–29 September 1999; pp. 322–327.

35. Anagnostopoulos, C.N.E.; Anagnostopoulos, I.E.; Loumos, V.; Kayafas, E. A license plate-recognition algorithm for intelligent transportation system applications. IEEE Trans. Intell. Transp. Syst. **2006**, 7, 377–392.

36. Hu, H.; Munoz, D.; Bagnell, J.A.; Hebert, M. Efficient 3-D scene analysis from streaming data. In Proceedings of IEEE International Conference on Robotics and Automation (ICRA), Karlsruhe, Germany, 6–10 May 2013.

37. Reed, I.S.; Yu, X. Adaptive multiple-band CFAR detection of an optical pattern with unknown spectral distribution. IEEE Trans. Acoust. Speech Signal Process. **1990**, 38, 1760–1770.

38. Chang, C.-I.; Chiang, S.-S. Anomaly detection and classification for hyperspectral imagery. IEEE Trans. Geosci. Remote Sens. **2002**, 40, 1314–1325.

39. Alonso, M.C.; Malpica, J.A. The combination of three statistical methods for visual inspection of anomalies in hyperspectral imageries. In Proceedings of 7th International Conference on Advances in Pattern Recognition (ICAPR), Kolkata, India, 4–6 February 2009; pp. 377–380.

40. Hongchao, F.; Wei, Y.; Long, T. Identifying man-made objects along urban road corridors from mobile LiDAR data. IEEE Geosci. Remote Sens. Lett. **2014**, 11, 950–954.

41. Kailath, T. The divergence and Bhattacharyya distance measures in signal selection. IEEE Trans. Commun. Technol. **1967**, 15, 52–60.

42. Puente, I.; González-Jorge, H.; Riveiro, B.; Arias, P. Accuracy verification of the Lynx mobile mapper system. Opt. Laser Technol. **2013**, 45, 578–586.

43. Heipke, C.; Mayer, H.; Wiedemann, C.; Jamet, O. Evaluation of automatic road extraction. Int. Arch. Photogramm. Remote Sens. **1997**, 32, 151–160.

44. Jensen, J.R.; Lulla, K. Introductory Digital Image Processing: A Remote Sensing Perspective, 2nd ed.; Prentice-Hall: Upper Saddle River, NJ, USA, 1996.

CHAPTER 10

Luminance-Corrected 3D Point Clouds for Road and Street Environments

Matti T. Vaaja [1,*], Matti Kurkela [1], Juho-Pekka Virtanen [1], Mikko Maksimainen [2], Hannu Hyyppä [1,3], Juha Hyyppä [4] and Eino Tetri [2]

[1] Department of Real Estate, Planning and Geoinformatics, Centre of Excellence in Laser Scanning Research (CoE-LaSR), Aalto University, FI-00076 Aalto, Finland

[2] Department of Electrical Engineering and Automation, Lighting Unit, Aalto University, FI-00076 Aalto, Finland

[3] Construction and Real Estate Hubic, Helsinki Metropolia University of Applied Sciences, FI-00079 Metropolia, Finland

[4] Finnish Geospatial Research Institute (FGI), Centre of Excellence in Laser Scanning Research (CoE-LaSR), Geodeetinrinne 2, FI-02430 Masala, Finland

ABSTRACT

A novel approach to evaluating night-time road and street environment lighting conditions through 3D point clouds is presented. The combination of luminance imaging and 3D point cloud acquired with a terrestrial laser scanner was used for analyzing 3D luminance on the road surface. A calculation of the luminance (cd/m^2) was based on the RGB output values of a Nikon D800E digital still camera. The camera was calibrated with a reference luminance source. The relative orientation between the luminance images and intensity image of the 3D point cloud was solved in order to integrate the data sets into the same coordinate system. As a result, the 3D model of road environment luminance is illustrated and the ability to exploit the method for evaluating the luminance distribution on the road surface is presented. Furthermore, the limitations and future prospects of the methodology are addressed. The method provides promising results for studying road lighting conditions in future lighting optimizations. The paper presents the methodology and its experimental application on a road section which consists of five luminaires installed on one side of a two-lane road in Otaniemi, Espoo, Finland.

KEYWORDS

road environment; lighting; 3D; point clouds; luminance; terrestrial laser scanning

1. INTRODUCTION

The function of road and street environment lighting is to maintain safety and comfort by providing acceptable visibility conditions during the night-time. Design of new road lighting setups, and retrofitting of old ones is affected by energy efficiency requirements [1] and by the ban of mercury lamps [2]. The increased awareness of mesopic vision may also affect the design of road lighting (e.g., [3,4]). The measurements of traffic lighting conditions are used to evaluate the performance of the lighting installation on road surfaces, road markings and furniture, or the glare from luminaires on surrounding areas can be investigated. Lighting measurements are also necessary for maintenance purposes. In the future, LED lights will be used more and more for illuminating road environments, because the luminous efficacies of the most recent LED luminaires are higher than the majority of the traditional light sources. Therefore, it is also important to evaluate the performance of LEDs for illuminating roadways.

With conventional gas discharge lamp luminaires, it is common to design outdoor lighting using a lumen maintenance factor of 80%. Lumen depreciation is due to lamp lumen depreciation but also due to luminaires getting dirty. A lumen maintenance factor of 80% is suitable if lamps are replaced with group replacement after three or four years, with burning hours thus being 12,000 to 16,000 hours. With LED luminaires the situation is different: their operating life time varies from 50,000 to 100,000 hours according to the manufacturer's information. However, the luminous flux of the LED chips is reduced during burning hours and the rate of depreciation is affected by the ambient temperature, for instance. In addition, air pollution effects on the lenses of the LED package and also on the external optics, thus reducing the luminous flux on road surface. Therefore, LED luminaires will bring about a new requirement for lighting measurements during burning hours in order to confirm that the LED lighting installation is still meeting the lighting needs after several thousand burning hours.

Road environment lighting conditions are evaluated using different methods, depending on the type of road or street: illuminance and its uniformity, luminance and its uniformity, glare restriction, target visibility, or color of the light. In this work, the lighting conditions are assessed in terms of luminance. Luminance describes the luminous intensity emitted from a particular area in a given direction. The SI unit for luminance is candela per square metre (cd/m^2). Luminance of the road is determined by the surface reflection properties and the light intensity based on the direction of light and the angle of reflection. Luminance measurements of road and street environments are needed to get data for analyzing the quality of road lighting and visibility conditions, for example.

Road lighting measurements can be carried out with spot luminance meters, which measure the luminance of a small (usually 1°) area at a time or by using an imaging luminance photometer (e.g., [5]). Zhou et al. [6] measured lighting illuminance levels for a moving vehicle and matched the measurements with the location data from a distance measurement instrument.

In recent years, laser scanning (LS) instruments have provided efficient and versatile applications for collecting 3D data for built and natural environments. Combined with positioning techniques LS produces the georeferenced point cloud, from which it is possible to calculate various surface, raster, and object models. In road and street environments,a mobile laser scanning (MLS) point cloud has been used for extracting different features, such as building façades, road surfaces, tunnels, railway tracks, poles, luminaires, and trees (e.g., [7,8,9,10,11,12,13,14,15]).

The fusion of 3D point cloud and image texture improves the interpretation and detection of features and objects. When considering data fusion, both 3D point clouds and 2D images must be registered in the common coordinate frame. The main approaches to registration include determining the sensor orientations separately using ground control features, solving the relative orientation between data sets, or making a system calibration of the integrated imaging system [16]. The combination of 3D point clouds with other sensor techniques can be found in a large number of published works. For example, point clouds have been integrated with data from a thermographic camera and ground-penetrating radar [17,18,19]. The result can also be a georeferenced panorama image, a textured surface, a building façade or a virtual environment [20,21,22,23]. In addition, point cloud rendering methods are used to improve the visualization (e.g., [24]).

The objective of this study is to demonstrate a novel approach to create luminance-corrected 3D point cloud data for evaluating night-time road lighting conditions and to address the limitations and future prospects.

2. STUDY AREA AND DATA ACQUISITION

2.1. Test Site

The research was conducted on a straight street section in Otaranta, Espoo, Finland (Figure 1). The experimental road section was approximately 150 m long and 6 m wide. The road has 1% slope in the longitudinal direction and 3% cross-section slope in both lanes. The difference between the lowest and highest point of the road section was 1 m. The road is surrounded by lawns and parking areas. The road section consists of five luminaires installed on one side of a two-lane road. The spacing between the poles was 32 m, with the luminaire being mounted at height of 10 m. Each luminaire had one 100 W High Pressure Sodium (HPS) light source. The road surface is classified as R3 in the CIE road surface classification, with a lighting class of AL4b, which is similar to ME4b in European standards, except for the longitudinal uniformity.

2.2. 3D Point Cloud

We used a Faro Focus 3D terrestrial laser scanner to produce a 3D point cloud of the Otaranta road. The Faro Focus 3D is a 905 nm, phase-based, continuous-wave laser scanner with a 305° × 360° field of view. The distance measurement accuracy is ±2 mm at a distance of 25 m [25]. The circular beam diameter at the exit is 3 mm. In total, we measured the Otaranta road from four separate scanning positions which were located between the lamp posts. 3D data was acquired with a resolution setting, providing a point spacing of 6 mm on an orthogonal target located some 10 m from the scanner. The objective of the laser scanning measurements was to obtain the comprehensive point coverage for the road surface and its surroundings. The TLS point clouds were co-registered using at least three sphere reference targets between every scan pair. The distance between the sphere target and scanner remained reasonable relative to the scanning resolution used. In this study, the distances varied from 5 to 15 m. The spheres were positioned with VRS-GNSS measurements. The overall quality of the registration on reference targets showed 1.0 cm mean error. The output of the TLS measurements was then georeferenced using the Finnish ETRS-TM35 coordinate system. In total, 73 million points were measured from the area.

2.3. Image Acquisition

Image acquisition was performed from the same locations using the same tripod setup as with TLS scans and for both directions of the road with a Nikon D800E camera (Figure 2). For the camera, the field of view was similar to the driver's point of view. The aperture of the camera was 5.6, exposure time 8 s and ISO value 100. These settings enabled luminance values of under 5 cd/m^2 with maximum dynamic range to be captured. All images were taken continuously under conditions where the existing luminance did not vary during the measurement. The photos were saved in NEF format (Nikon's lossless RAW file format) to record light linearly and maximize image quality for post-processing.

To address a 3D point with an absolute luminance value, the absolute luminance value must first be clarified. A digital camera can be used as an imaging luminance photometer [26,27,28]. To interpret the images taken as matrices, the absolute luminance values of the pixel in the raw image of the camera used needed to be calibrated. In addition, the camera was geometrically calibrated and lens distortions were corrected by applying Brown's model [29].

Luminance Calibration

For obtaining luminance values from digital images, Equations (1) and (2) are applied. The relation between aperture, exposure time, ISO, and luminance is explained in the following Equation [28]:

$$K = \frac{N_d f_s^2}{L_s t S_{ISO}} \quad (1)$$

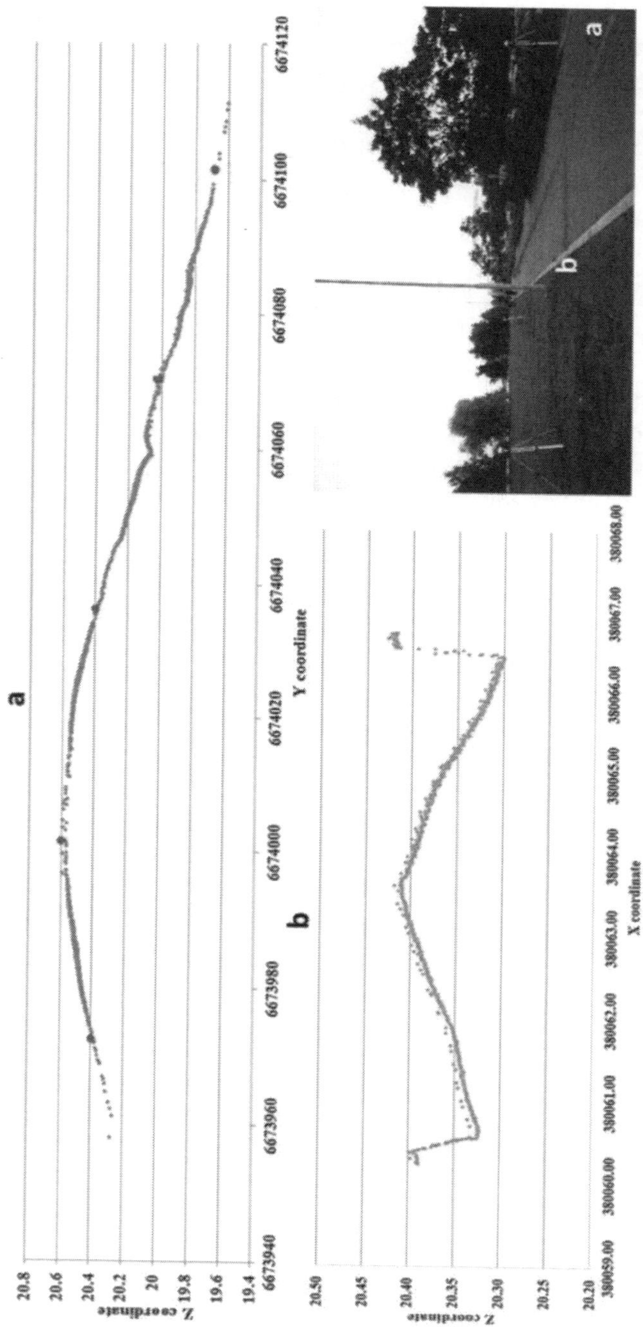

Figure 1. Profile view of the Otaranta road center line (**a**) and cross-section (**b**). Z coordinates are presented in GRS80 (Geodetic Reference System 1980), Y and X coordinates in ETRS-TM35FIN, units expressed in meters (positions of poles are marked with red dots in picture a).

where

- K is the calibration constants
- N_d is the digital value Y of the pixel in the raw image obtained with equation 2,
- f_s is the aperture,
- L_s is the luminance of the scene in cd/m^2
- t is the exposure time in seconds and
- S_{ISO} is the ISO value.

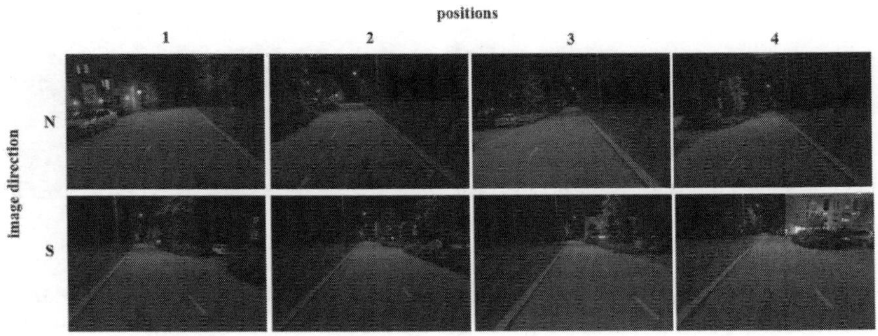

Figure 2. The image data included eight images covering a 150-metre long road section under night-time lighting conditions. The images were taken from four different positions and in both directions (north and south).

The value of Y in the CIE XYZ color space [30] is equivalent to the photopic luminance value [31]. The correlation between the RGB color space used in the Nikon D800E camera and Y in CIE XYZ is defined by the IEC standard [32]:

$$Y = 0.2162R + 0.7152G + 0.0722B \quad (2)$$

To determine the ratio of absolute and relative luminances with specific exposure settings, a set of images with known luminance were acquired. An Optronic Laboratories Inc. model 455-6-1 reference luminance source (Figure 3a) was used to create uniform circular luminance areas. To obtain low luminance values corresponding to night-time road conditions, a neutral density (ND) filter was used in front of the opening of the luminance source's integrating sphere. The transfer ratio of the ND filter was measured using an LMK Mobile Advanced imaging luminance photometer, and found to be 1/41.8. After this, a series of 54 images of the ND-filtered luminance source were taken with the Nikon D800E. The calibration was performed with the same exposure settings as used in the field (aperture of the lens 5.6, exposure time of 8 s and the ISO value 100). The luminance source was adjusted to 54 different luminance ranging from 0.0031 to 23.95 cd/m^2. The median RGB channel values from an area of 100 × 100 pixels in the center of the image (Figure 3b) covering the opening of the luminance source were used in calibrating the camera.

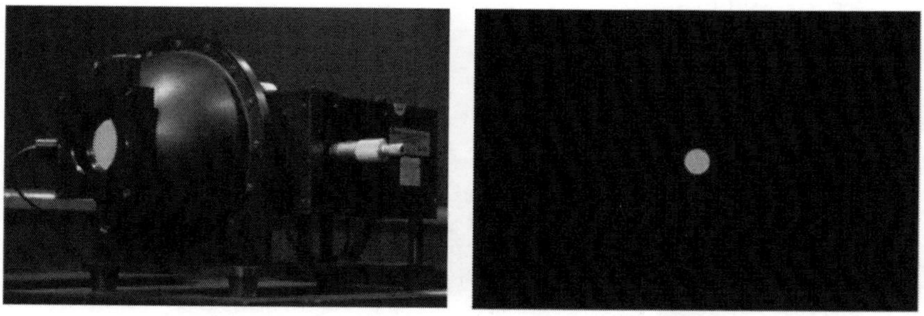

(a) (b)

Figure 3. (**a**) Optronic Laboratories Inc. model 455-6-1 reference luminance source used for calibration. (**b**) The image covering the opening of the luminance source.

Using the transformation function (Equation (2)) with an RGB image, a matrix of relative luminance values was obtained. With the absolute luminance value of the image being known, the linear ratio between relative and absolute luminance values was identified. It was found that with the exposure settings used, the maximum measurable luminance value was 2.9 cd/m^2 because, with greater values, one of the RGB channels reached its limits. The ratio and the RGB to Y function (Equation (2)) can now be used to interpret every image taken with the calibrated camera, using the same exposure settings, as a matrix of absolute luminance values.

3. INTEGRATION OF LUMINANCE VALUES AND 3D POINT CLOUD

Our approach to derive a 3D luminance model for analyzing road environment lighting conditions is based on the fusion of TLS measurements and luminance values derived from digital images. The creation of luminance colored 3D point clouds requires that images and laser scanning data are in the same coordinate system. We integrated these terrestrial data sources by determining the orientation of undistorted RGB images relative to intensity images of individual terrestrial laser scans (Figure 4). We identified a set of corresponding points for every image pair by using recognizable road features such as paintings and kerbstones. The results of the registration showed that the mean distance between the 2D projection of the selected 3D points and the corresponding points that have been selected in the RGB pictures for all image pair was 4.7 pixels. The uncertainty was mainly caused by lower resolution on TLS intensity images. After determining the orientation, the images were replaced by calculated luminance images (see Section 2.3.1) in order to derive 3D luminance point clouds. The RGB images were used for orientation for the reason that they allowed us to observe the road features more clearly.

3.1. 3D Luminance Point Cloud

Once all images have been registered with the corresponding TLS intensity images, the 3D luminance data of the road can be generated. In other words, every measurement point contains x, y, and z coordinates and a calibrated luminance value. The output is that individual TLS scans are textured with two images covering both directions of the road. That enabled us to obtain comprehensive point coverage for the studied road surface. The perpendicular directions along the road were not evaluated in this study. The output is presented in Figure 5.

3.2. Statistical Analysis of Road Surface Luminance

The generated 3D luminance data enables us to evaluate the values according to their position or location on the road. Figure 6 shows the scatter plot of the right and left lane luminance values in the longitudinal direction along the centerline of each lane. The luminance values were sampled by calculating an average value for 10 × 10 cm areas for creating the scatter plots in Figure 6 and Figure 7.

Figure 4. Road features such as paintings and kerbstones were used as corresponding points between a TLS-based intensity image and an RGB image for resolving the relative orientation.

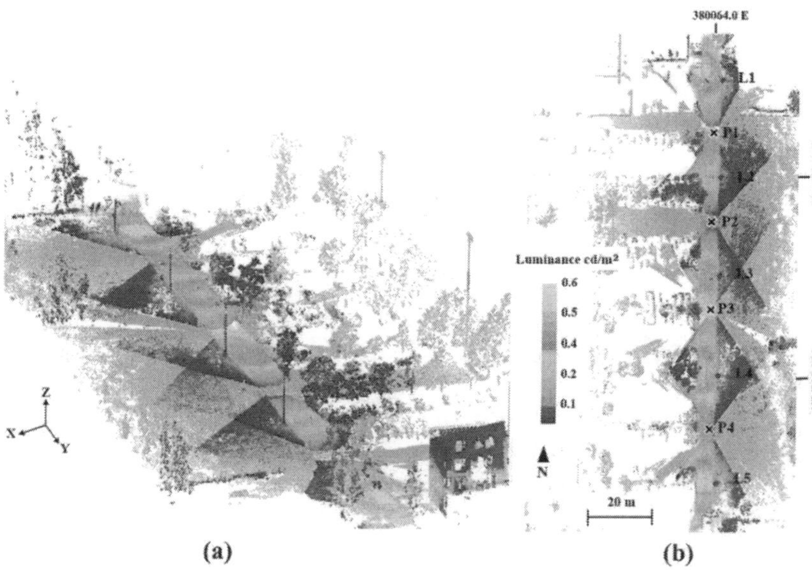

Figure 5. The derived 3D luminance data of the road section; (**a**) bird's-eye view and (**b**) orthographic view, from above. The points which are not textured by images are shown in grey. The positions of the scan and image stations are marked with black crosses (P1–P4) and lamp posts are marked with red dots (L1–L5). Coordinates are presented in ETRS-TM35FIN.

Figure 7 shows the scatter plot for comparison between the right and left lane luminance values captured from different directions (positions 3 and 4). The right and left lane samples cover a 20 × 0.1 m area next to the lamp pole and in the middle of the lane. The results show equal luminance values for different capturing directions and they confirm the diffuse reflection behavior of the studied road. We found the standard deviation for the right lane observations to be 0.029 cd/m² and for the left lane 0.036 cd/m².

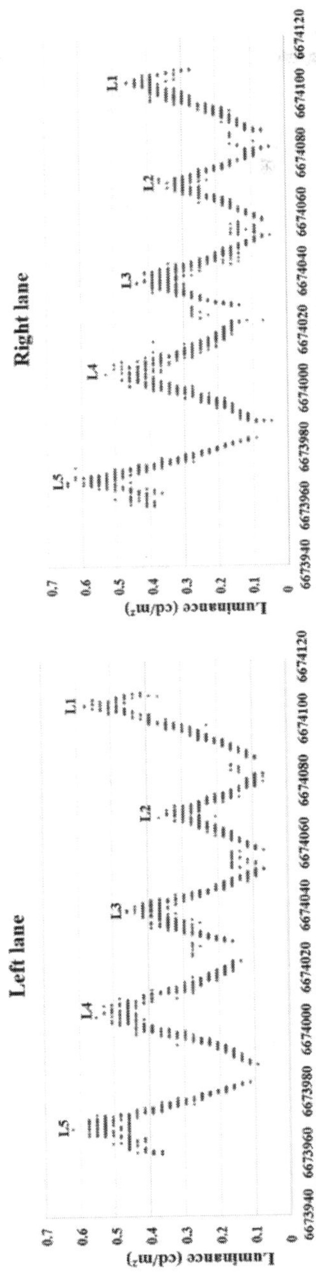

Figure 6. The **(left)** and **(right)** lane luminance values in the longitudinal direction along the centreline of each lane. The position of the lamps (L1–L5) are marked above the corresponding luminance peaks.

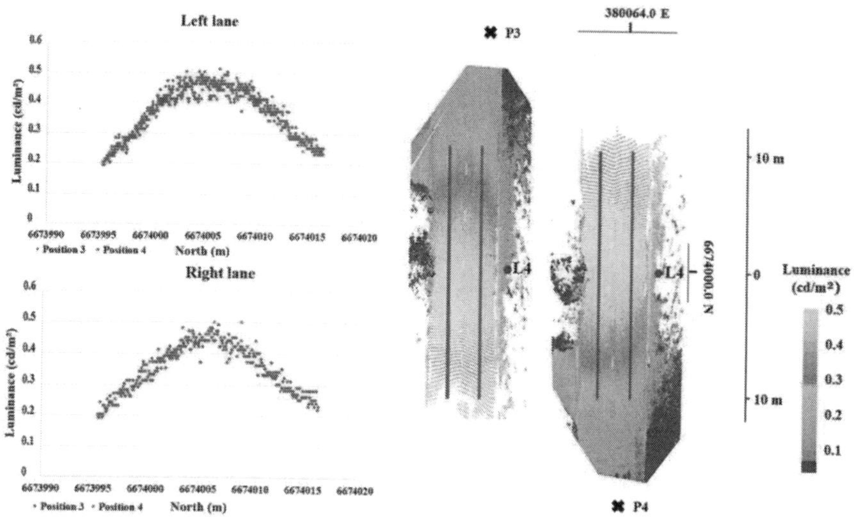

Figure 7. The **(left)** and **(right)** lane luminance values in the longitudinal direction along the centreline of each lane. The scatter plot shows the values captured from different directions (position 3, P3; and position 4, P4). The position of the lamp pole and samples are marked in red and blue. Coordinates are presented in ETRS-TM35FIN.

Further, the measured road surface luminances are presented in Table 1. The road surface next to lamp poles was measured with TLS and luminance imaging from both directions of the road as shown in Figure 7. Table 1 compares the results from northerly and southerly directions. Before comparison, the luminance values were sampled by calculating an average value for 10 × 10 cm areas. We derived Average luminance (L_m), Overall Uniformity of luminance (U_o), and Longitudinal Uniformity (U_l) values for three different sections of the road next to lamp poles 2, 3, and 4 [33]. L_m is the average luminance of the carriageway measured at the center of each lane; U_o is the ratio of minimum to average luminance in the defined road area, and U_l refers to the ratio of minimum to maximum luminance in the longitudinal direction along the centerline of each lane. In Table 1 L_m and U_l value is calculated from a 20 × 0.1 m area next to the lamp pole. U_o value is calculated from a 20 × 2.0 m area next to the lamp pole.

Table 1. The comparison of the road surface luminance measured from northerly and southerly directions (**N** and **S**). Average luminance (L_m), Overall Uniformity of luminance (U_o), and Longitudinal Uniformity (U_l) values were derived for three different sections of the road next to lamp poles 2, 3, and 4. L_m and U_l are calculated separately for the left and right lane.

Lamp	Direction (/Position)	L_m, Left Lane (cd/m²)	L_m, Right Lane (cd/m²)	U_o (%)	U_l, Left Lane (%)	U_l, Right Lane (%)
L2	N (P1)	0.211	0.221	0.425	0.286	0.267
	S (P2)	0.206	0.214	0.433	0.385	0.286
L3	N (P2)	0.316	0.307	0.446	0.368	0.389
	S (P3)	0.298	0.280	0.396	0.389	0.353
L4	N (P3)	0.383	0.369	0.495	0.478	0.500
	S (P4)	0.346	0.346	0.399	0.318	0.364

4. DISCUSSION

The integration of TLS and luminance imaging in this study proved to be a useful method for creating luminance-corrected 3D point cloud data for evaluating night-time road lighting conditions. In other words, the presented method combines calibrated luminance values (cd/m²) with 3D points. In other studies, the term "luminance correction" has also been connected with a point cloud rendering approach which removes luminance inconsistencies caused by different lighting conditions when taking the various 3D scans with color information [34].

The presented method requires that the relative orientation between the luminance images and intensity images of the 3D point cloud is solved in order to integrate the data sets into the same coordinate system. The registration error for all image pairs was 4.7 pixels, which was mainly caused by lower resolution on TLS intensity images. The calculation of the luminance was based on the RGB output values of a Nikon D800E digital still camera. The camera was calibrated with a reference luminance source. In terms of luminance imaging, it should be noted that the intensity of the reflected light may vary according to the viewing direction, particularly if the road surface is wet. We confirmed the diffuse reflection of the studied road by comparing the luminance values captured from different directions, and we achieved relevant results when the imaging was performed from a driver's point of view (Figure 7). The standard deviations of the comparison were 0.029 cd/m² and 0.036 cd/m². In this study, we focused on measuring the lighting conditions on the road surface. However, the method can also be used to evaluate the glare in the surrounding area. Luminance-corrected 3D point cloud could also be combined with eye-tracking data [4]. A detailed 3D point cloud produced by TLS also allows for the detection of the shadowing objects, poles, pole orientations, traffic signs, road markings, information boards, and other objects along the road. The results also confirmed that the method can be used to distinguish the variation in the lighting conditions on different sections of the road (Figure 6). Hence, the method is a new tool for monitoring lighting performance and variations of visibility level along the road (see e.g., [35]). In addition, road lighting standards set requirements for maintained average luminance and its uniformity [33]. These values have been calculated in Table 1.

The TLS method requires that the road is closed while the measurements are being taken. In this study, the TLS and image data acquisition from one measurement station took around 10 min; this limits the size of the study area to be examined. This constraint can be avoided by integrating luminance imaging with mobile laser scanning (MLS). This technology is a novel method for collecting street or road inventory data for design, maintenance, rehabilitation, and environment purposes. The MLS method has the ability to identify objects and areas that are difficult and time-consuming to identify via other mapping methods. The benefits of MLS are apparent in a measuring rate context, especially when compared to the TLS method. A good and recent comparison of performance of MLS systems on the market has been made by Kaartinen et al. [36]. In this study, the ROAMER system is compared with commercial systems from Riegl (Riegl VMX-250), Optech (Optech Lynx Mobile Mapper), and Streetmapper 360. In tests performed in a real life situation in the Espoonlahti area, the systems could reach an elevation accuracy of 35 mm from a range of 35 m, with the best system being capable of a planimetric accuracy of 25 mm from a range of 45 m [36]. When integrated to MLS system, the luminance mapping of roads could be performed parallel to MLS road inventories. The requirement for systems capable of measuring road lighting of long road segments, and producing georeferenced data has been brought up in research literature [6,37]. Integration of MLS and luminance measurement could also be applied for this purpose. In addition, other analyses concerning the road environment could be performed from the same data set, as mentioned above.

In this study, the photopic transformation function, presented in Equation (2), was used to calculate luminance values from an RGB image. Related to this, there are two main types of photoreceptor cell in the human eye: cone cells and rod cells. Cone cells are used when there is plenty of light. They are located in the fovea—the central area of the retina—and they make acute color vision possible. Rod cells are specialized in dark vision and are unable to distinguish between colors. Vision where only cone cells are used is called photopic vision, whereas in scotopic vision, only rod cells are used. In mesopic vision, both cone and rod cells are used simultaneously. It is debatable how well the photopic transformation function presented in Equation (2) actually performs. The actual transformation function is different in each digital camera and may differ more or less from the IEC standard [32]. For low scotopic luminance values (<0.005 cd/m^2), the transformation function is also different and individual for each digital camera. In addition, if both photopic and scotopic transfer functions are known, it is possible to obtain the mesopic luminance value using the iterative formula presented in CIE 191 [38]. This complex extension was beyond the scope of this study but it could be considered in the future.

5. CONCLUSIONS

A promising and novel 3D luminance evaluation approach of night-time road lighting conditions based on luminance-corrected 3D point clouds is demonstrated in this paper. Despite developments in laser scanning technology and applications, we have not found any other studies that have implemented the fusion of laser scanning and luminance imaging data and created luminance-corrected 3D point cloud data. A detailed road environment model with luminance values is obtained using terrestrial laser scanning and luminance imaging with a Nikon D800E camera. The luminance values of the RGB images were calibrated with a reference luminance source, and the calibration showed a linear ratio between the digital values of the Nikon D800E camera and the reference luminance. We integrated these terrestrial data sources in the same coordinate system by determining the orientation of undistorted RGB images relative to the intensity images of individual terrestrial laser scans. In the achieved 3D point cloud every measurement point had also its own calibrated luminance value in addition to coordinates. We assume that in the future, road lighting conditions can also be evaluated by integrating luminance imaging with mobile laser scanning, which improves the measurement rates and increases cost effectiveness.

ACKNOWLEDGMENTS

This research project was supported by the Academy of Finland, the Centre of Excellence in Laser Scanning Research (CoE-LaSR) (272195), the Aalto Energy Efficiency Research Programme (Light Energy—Efficient and Safe

Traffic Environments project), the EUE project (2141226), ERDF (projects “AKAI”, 301130 and "Soludus", 301192), the Aalto University doctoral program and city of Espoo.

AUTHOR CONTRIBUTIONS

Matti Vaaja was the main author of the article and was responsible for the TLS data acquisition, data integration and analysis. Matti Kurkela was responsible for image acquisition. Mikko Maksimainen, Juho-Pekka Virtanen, Matti Kurkela and Matti Vaaja implemented the luminance calibration. The article was improved by the contributions of all the co-authors at various stages of the analysis and writing process.

CONFLICTS OF INTEREST

The authors declare no conflict of interest.

REFERENCES

1. Kostic, M.; Djokic, L. Recommendations for energy efficient and visually acceptable street lighting. Energy **2009**, 34, 1565–1572.

2. EC 245/2009. COMMISSION REGULATION (EC) No 245/2009 of 18 March 2009 implementing Directive 2005/32/EC of the European Parliament and of the Council with regard to ecodesign requirements for fluorescent lamps without integrated ballast, for high intensity discharge lamps, and for ballasts and luminaires able to operate such lamps, and repealing Directive 2000/55/EC of the European Parliament and of the Council. Available online: http://eur-lex.europa.eu/LexUriServ/ LexUri Serv.do?uri=OJ:L:2009:076:0017:0044:en:PDF (accessed on 2 September 2015).

3. Kostic, M.; Djokic, L.; Pojatar, D.; Strbac-Hadzibegovic, N. Technical and economic analysis of road lighting solutions based on mesopic vision. Build. Environ. **2009**, 44, 66–75.

4. Cengiz, C.; Kotkanen, H.; Puolakka, M.; Lappi, O.; Lehtonen, E.; Halonen, L.; Summala, H. Combined eye-tracking and luminance measurements while driving on a rural road: Towards determining mesopic adaptation luminance. Light. Res. Technol. **2013**.

5. Ekrias, A.; Eloholma, M.; Halonen, L.; Song, X.J.; Zhang, X.; Wen, Y. Road lighting and headlights: Luminance measurements and automobile lighting simulations. Build. Environ. **2008**, 43, 530–536.

6. Zhou, H.; Pirinccioglu, F.; Hsu, P. A new roadway lighting measurement system. Transp. Res. Part C Emerg. Technol. **2009**, 17, 274–284.

7. Manandhar, D.; Shibasaki, R. Auto-extraction of urban features from vehicle-borne laser data. Int. Arch. Photogramm. Remote Sens. Spat. Inf. Sci. **2012**, 34, 650–655.

8. Jochem, A.; Höfle, B.; Rutzinger, M. Extraction of vertical walls from mobile laser scanning data for solar potential assessment. Remote Sens. **2011**, 3, 650–667.

9. Jaakkola, A.; Hyyppä, J.; Hyyppä, H.; Kukko, A. Retrieval algorithms for road surface modelling using Laser-based mobile mapping. Sensors **2008**, 8, 5238–5249.

10. Puente, I.; González-Jorge, H.; Martínez-Sánchez, J.; Arias, P. Automatic detection of road tunnel luminaires using a mobile LiDAR system. Measurement **2014**, 47, 569–575.

11. Lehtomäki, M.; Jaakkola, A.; Hyyppä, J.; Kukko, A.; Kaartinen, H. Detection of vertical pole-like objects in a road environment using vehicle-based laser scanning data. Remote Sens. **2010**, 2, 641–664.

12. Rutzinger, M.; Pratihast, A.K.; Elberink, S.O.; Vosselman, G. Detection and modeling of 3D trees from mobile laser scanning data. Int. Arch. Photogramm. Remote Sens. Spat. Inf. Sci. **2010**, 38, 520–525.

13. Yang, B.; Wei, Z.; Li, Q.; Li, J. Automated extraction of street-scene objects from mobile Lidar point clouds. Int. J. Remote Sens. **2012**, 33, 5839–5861.

14. Arastounia, M.; Elberink, S.O.; Khoshelham, K.; Diaz Benito, D. Rail track detection and modeling in mobile laser scanner data. ISPRS Ann. Photogram. Remote Sens. Spat. Inform. Sci. **2013**, 5, 223–228.

15. Wu, B.; Yu, B.; Yue, W.; Shu, S.; Tan, W.; Hu, C.; Huang, Y.; Wu, J.; Liu, H. A voxel-based method for automated identification and

morphological parameters estimation of individual street trees from mobile laser scanning data. Remote Sens. **2013**, 5, 584–611.

16. Rönnholm, P.; Honkavaara, E.; Litkey, P.; Hyyppä, H.; Hyyppä, J. Integration of laser scanning and photogrammetry (Key-note). Int. Arch. Photogramm. Remote Sens. Spat. Inf. Sci. **2007**, 36, 355–362.
17. Alba, M.I.; Barazzetti, L.; Scaioni, M.; Rosina, E.; Previtali, M. Mapping infrared data on terrestrial laser scanning 3D models of buildings. Remote Sens. **2011**, 3, 1847–1870.
18. González-Aguilera, D.; Rodriguez-Gonzalvez, P.; Armesto, J.; Lagüela, S. Novel approach to 3D thermography and energy efficiency evaluation. Energy Build. **2012**, 54, 436–443.
19. Lubowiecka, I.; Arias, P.; Riveiro, B.; Solla, M. Multidisciplinary approach to the assessment of historic structures based on the case of a masonry bridge in Galicia (Spain). Comput. Struct. **2011**, 89, 1615–1627.
20. Zhu, L.; Hyyppä, J.; Kukko, A.; Jaakkola, A.; Lehtomäki, M.; Kaartinen, H.; Chen, R.; Pei, L.; Chen, Y.; Hyyppä, H.; Rönnholm, P.; Haggren, H. 3D city model for mobile phone using MMS data. In Proceedings of the Urban Remote Sensing Joint Event, Shanghai, China, 20–22 May 2009.
21. Alamús, R.; Kornus, W. DMC geometry analysis and virtual image characterisation. Photogramm. Record **2008**, 23, 353–371.
22. Brenner, C.; Haala, N. Rapid acquisition of virtual reality city models from multiple data sources. Int. Arch. Photogramm. Remot. Sens. **1998**, 32, 323–330.
23. Virtanen, J.-P.; Puustinen, T.; Pennanen, K.; Vaaja, M.T.; Kurkela, M.; Viitanen, K.; Rönnholm, P. Customized visualizations of urban infill development scenarios for local stakeholders. J. Build. Constr. Plan. Res. **2015**, 3, 68–81.
24. Kanzok, T.; Linsen, L.; Rosenthal, P. An interactive visualization system for huge architectural Laser Scans. In Proceedings of the 10th International Conference on Computer Graphics Theory and Applications, Berlin, Germany, 11–14 March 2015; pp. 265–273.
25. Chow, J.; Lichti, D.; Teskey, W. Accuracy assessment of the Faro Focus3D and Leica HDS6100 panoramic type terrestrial laser scanner through point-based and plane-based user self-calibration. In Proceedings of the FIG Working Week: Knowing to Manage the

Territory, Protect the Environment, Evaluate the Cultural Heritage, Rome, Italy, 6–10 May 2012.

26. Meyer, J.E.; Gibbons, R.B.; Edvards, C.J. Development and Validation of a Luminance Camera, Final Report; Virginia Tech Transportation Institute: Blacksburg, VA, USA, 2009.

27. Wüller, D.; Gabele, H. The usage of digital cameras as luminance meters. Proc. SPIE **2007**.

28. Hiscocks, P.D.; Eng, P. Measuring Luminance with a Digital Camera: Case History. Available online: http://www.ee.ryerson.ca: 8080/~phiscock/astronomy/light-pollution/luminance-case-history.pdf (accessed on 2 September 2015).

29. Fryer, J.G.; Brown, D.C. Lens distortion for close-range photogrammetry. Photogramm. Eng. Remote Sens. **1986**, 52, 51–58.

30. CIE 1932. In Commission Internationale de l'Eclairage Proceedings; Cambridge University Press: Cambridge, UK, 1931.

31. CIE 1926. In Commission Internationale de l'Eclairage Proceedings; Cambridge University Press: Cambridge, UK, 1924.

32. IEC 61966-2-1. In Multimedia Systems and Equipment—Colour Measurements and Management—Part 2-1: Colour Management—Default RGB Color Space—sRGB; International Electrotechnical Commission: Geneva, Switzerland, 1999.

33. EN 13201. European Standard EN 13201-2. Road lighting. Part 2: Performance Requirements. Available online: https://www. fer. unizg.hr/_download/repository/en_13201-2_.pdf (accessed on 2 September 2015).

34. Kanzok, T.; Linsen, L.; Rosenthal, P. On-the-fly luminance correction for rendering of inconsistently lit point clouds. J. WSCG **2012**, 20, 161–169.

35. Güler, Ö.; Onaygil, S. The effect of luminance uniformity on visibility level in road lighting. Light. Res. Technol. **2003**, 35, 199–213.

36. Kaartinen, H.; Hyyppä, J.; Kukko, A.; Jaakkola, A.; Hyyppä, H. Benchmarking the performance of mobile laser scanning systems using a permanent test field. Sensors **2012**, 12, 12814–12835.

37. Cai, H.; Li, L. Measuring light and geometry data of roadway environments with a camera. J. Transp. Technol. **2014**, 4, 44–62. [

38. Commission Internationale de l'Eclairage. Recommended System for Mesopic Photometry Based on Visual Performance; CIE Publication 191: Vienna, Austria, 2010; p. 79.

Index

S

T